INTRODUCTORY
TOPOLOGY

Exercises and Solutions

INTRODUCTORY
TOPOLOGY

Exercises and Solutions

Mohammed Hichem Mortad

University of Oran, Algeria

 **World Scientific**

NEW JERSEY · LONDON · SINGAPORE · BEIJING · SHANGHAI · HONG KONG · TAIPEI · CHENNAI

Published by

World Scientific Publishing Co. Pte. Ltd.

5 Toh Tuck Link, Singapore 596224

USA office: 27 Warren Street, Suite 401-402, Hackensack, NJ 07601

UK office: 57 Shelton Street, Covent Garden, London WC2H 9HE

Library of Congress Cataloging-in-Publication Data
Mortad, Mohammed Hichem, 1978– author.
   Introductory topology : exercises and solutions / by Mohammed Hichem Mortad (University of
Oran, Algeria).
     pages cm
   Includes bibliographical references and index.
   ISBN 978-9814583817 (pbk : alk. paper)
   1. Topology--Problems, exercises, etc. I. Title.
   QA611.M677 2014
   514.076--dc23
                          2014003079

British Library Cataloguing-in-Publication Data
A catalogue record for this book is available from the British Library.

Printed in Singapore by World Scientific Printers.

Preface

Topology is a major area in mathematics. At an undergraduate level, at a research level, and in many areas of mathematics (and even outside mathematics in some cases), a good understanding of the basics of the theory of general topology is required. Many students find the course "Topology" (at least in the beginning) a bit confusing and not too easy (even hard for some of them) to assimilate. It is like moving to a different place where the habits are not as they used to be, but in the end we know that we have to live there and get used to it. They are usually quite familiar with the real line $\mathbb{R}$ and its properties. So when they study topology they start to realize that not everything true in $\mathbb{R}$ needs to remain true in an arbitrary topological space. For instance, there are convergent sequences which have more than one limit, the identity mapping is not always continuous, a normally convergent series need not converge (although the latter is not within the scope of this book). So in many references, they use the word "usual topology of $\mathbb{R}$", a topology in which things are as usual! while there are many other "unusual topologies" where things are not so "usual"!

The present book offers a good introduction to basic general topology throughout solved exercises and one of the main aims is to make the understanding of topology an easy task to students by proposing many different and interesting exercises with very detailed solutions, something that it is not easy to find in another manuscript on the same subject in the existing literature. Nevertheless, and in order that this books gives its fruits, we do advise the reader (mainly the students) to use the book in a clever way inasmuch as while the best way to learn mathematics is by doing exercises, the worst way of doing exercises is to read the solution without thinking about how to solve the exercises (at least for some time). Accordingly, we strongly recommend the student to attempt the exercises before consulting the solutions. As a Chinese proverb says: If you give someone a fish, then you have given him to eat for one day, but if you teach him fishing, then you have given him food for everyday. So we hope the students are going to learn to "fish" using this book.

The present manuscript is mainly intended for an undergraduate course in general topology. It does not include algebraic and geometric topologies. Other topics such as: nets, topologies of infinite products, quotient topology, first countability, second countability and the T_i separation axioms with $i = 0, 3, 4, 5$ are not considered either or are not given much attention. It can be used by students as well as lecturers and anyone who needs the basic tools of topology. Teaching this course several times with many different exercises each year has allowed me to collect all the exercises given in this book. I relied on many references (I cannot remember all of them but most of them can be found in the bibliography) in lectures and tutorials. If there is some source which I have forgotten to mention, then I sincerely apologize for that.

Let us now say a few words about the contents of this book. The exercises on the subjects covered in this book can be used for a one semester course of 14 weeks.

The book is divided into two parts. In Part 1, each chapter (except for the first one) contains five sections. They are:

(1) What You Need to Know: In this part, we briefly recall the essential of notions and results which are needed for the exercises. No proofs are given. We just note that this part cannot in any case replace a detailed course on the subject. The reader may also wish to consult the following references for further reading: [1], [2], [3], [6], [7], [9], [10], [11], [13], [14], [15] and [16].

(2) True or False: In this part some interesting questions are proposed to the reader. They also contain common errors which appear with different students almost every year. Thanks to this section, students should hopefully avoid making many silly mistakes. This part is an important back-up for the "What You Need to Know" section. Readers may even find some redundancy, but this is mainly because it is meant to test their understanding.

(3) Exercises with Solutions: The major part and the core of each chapter where many exercises are given with detailed solutions.

(4) Tests: This section contains short questions given with just answers or simply hints.

(5) More Exercises: In this part some unsolved exercises are proposed to the interested reader.

In Part 2, the reader finds answers to the questions appearing in the section "True or False" as well as solutions to Exercises and Tests.

The prerequisites to use this book are basics of: functions of one variable (some of several variables calculus is also welcome though not very much), sequences and series, and set theory.

Since the terminology in topology is rich and may be different from a book to another, we do encourage the readers to have a look at the "Notations and Terminology" chapter to avoid an eventual confusion or ambiguity with symbols and notations.

Before finishing, I welcome and I will be pleased to receive any suggestions, questions (as well as pointing out eventual errors and typos) from readers at my email: **mhmortad@gmail.com.**

Last but not least, thanks are due in particular to Dr Lim Swee Cheng and Ms Tan Rok Ting, and all the staff of World Scientific Publishing Company for their patience and help.

Oran on September the 24th, 2013
Mohammed Hichem Mortad
Department of Mathematics
Faculty of Exact and Applied Sciences
The University of Oran (Algeria)

Contents

Notation and Terminology

0.1. Notation

- $\mathbb{N}$ is the set of natural numbers, i.e. $\mathbb{N}$ is the set $\{1, 2, \cdots\}$ (note that $\mathbb{N}$ in the French literature contains 0 too).
- $\mathbb{Z}$ the set of all integers while $\mathbb{Z}^+$ is the set of positive numbers.
- $\mathbb{Q}$ the set of rational numbers.
- $\mathbb{R}$ the set of real numbers.
- $\mathbb{C}$ the set of complex numbers.
- $[a, b]$ the closed interval with endpoints a and b.
- (a, b) the closed interval with endpoints a and b.
- (a, b) also denotes an ordered pair.
- i the complex square root of -1.
- $x \mapsto e^x$ the usual exponential function.
- The complement of a set A in X is often denoted by A^c and occasionally it is denoted by $X \setminus A$.
- The empty set is denoted by $\varnothing$.
- $d(x, A)$ is the distance between a point x in X to a set $A \subset X$ where (X, d) is a metric space.
- $d(A)$ is the diameter of a set A in a metric space (X, d).
- In a metric space (X, d), the open ball of radius $r > 0$ and center $x \in X$ is denoted by $B(x, r)$.
- In a metric space (X, d), the closed ball of radius $r > 0$ and center $x \in X$ is denoted by $B_c(x, r)$.
- In a metric space (X, d), the sphere of radius $r > 0$ and center $x \in X$ is denoted by $S(x, r)$.

- The interior of a set A in a topological space is denoted by $\overset{\circ}{A}$.
- The closure of a set A in a topological space is denoted by $\overline{A}$.
- A' denotes the derived set of A, that is the set of limit points of A (where A is a subset of a topological space X).
- The frontier of a set is denoted by Fr.
- card denotes the cardinal number of a set.
- $\mathbb{R}_\ell$ is the lower limit topology.
- $\mathbb{R}_K$ is the K-topology.

- The set of neighborhoods of a point $x \in X$, where X is a topological space, is denoted by $\mathcal{V}(x)$.
- C_x denotes the (connected) component of $x \in X$ where X is a topological space.
- The real part of a complex number is denoted by Re.
- The imaginary part of a complex number is denoted by Im.
- The conjugate of a complex number z is denoted by $\bar{z}$.
- $C([0,1])$ is the space of real-valued continuous functions on $[0,1]$ taking values in $\mathbb{R}$. If the field of values is $\mathbb{C}$, then this will be clearly mentioned.
- $C^1([0,1])$ is the space of real-valued continuous functions defined on $[0,1]$, differentiable and having a continuous derivative.
- $(\mathbb{R}, |\cdot|)$ is $\mathbb{R}$ equipped with the standard or the usual metric, i.e. the absolute value function.

0.2. Terminology

- There is a general comment which should be remembered by the reader. *Except* in the sections "True or False", if we say $\mathbb{R}$ without specifying the topology or the metric, then this means $\mathbb{R}$ endowed with its usual topology or metric. This applies to $\mathbb{C}$, $\mathbb{R}^n$ and $\mathbb{C}^n$ too.
- It will be comprehensible from the context whether "$\subset$" is for comparing two sets or two topologies.
- It will be clear from the context whether (a, b) is the ordered pair or the open interval.
- When it is not too important to specify the metric or the topology, we simply say the topological (or metric) space X.
- From time to time the reader will see "(why?)". In such a case, this means that this a question whose answer should be known by the reader. This is used by other authors such as J. B. Conway (see [4]) and also M. Stoll (see [15]). Other expressions such as "(is it not?)" are also used.
- A clopen set is a set that it is closed and open simultaneously.
- The letters "w.r.t." stand for with respect to.
- As it is used almost everywhere, "iff" means if and only if (for the fun, the French use "ssi" for "si et seulement si". Even in Arabic, it has been sorted out by doubling a letter in the end!).
- WLOG, as it pleases many, stands for "without loss of generality".

Part 1

Exercises

CHAPTER 1

General Notions: Sets, Functions et al

1.1. What You Need to Know

In this section, we assume the reader is familiar with basic real analysis including notions and results on sets and functions among others. So we only recall results on countability.

DEFINITION. *A set A is called **finite** if it is empty or if there is a bijection $f : A \to \{1, 2, \cdots, n\}$. A set is said to be **infinite** if it is not finite.*

THEOREM. *If a set A is finite, then there is no bijection of A with any proper subset of A.*

EXAMPLE. *The set $\mathbb{N}$ is infinite.*

DEFINITION. *A set A is called **countable** (or **denumerable**) if it is finite or if there is a bijection $f : A \to \mathbb{N}$. A set which is not countable is called **uncountable**.*

EXAMPLE.
  (1) *The set of even integers is countable.*
  (2) *$\mathbb{Z}$ is countable.*

When looking for a bijection between $\mathbb{N}$ and a given set A, it may happen that we find a function f which is injective but not surjective and vice versa. The next result is therefore quite practical

THEOREM. *A non-empty set A is countable iff one of the following holds:*
  (1) *There exists an injective map $f : A \to \mathbb{N}$;*
  (2) *There exists a surjective map $g : \mathbb{N} \to A$.*

PROPOSITION. *If X is countable, then so is any $A \subset X$.*

THEOREM.
  (1) *A countable union of countable sets remains countable.*
  (2) *A finite product of countable sets is also countable.*

THEOREM. *Let A be a non-empty set, and let $\mathcal{P}(A)$ be the set of all subsets of A. Then there si no injective map $f : \mathcal{P}(A) \to A$*

COROLLARY. *$\mathcal{P}(\mathbb{N})$ is uncountable.*

We finish this section with an important set in analysis. To define it, consider first the set $A_0 = [0,1]$. The set A_1 is obtained from A_0 be removing the middle third interval $(\frac{1}{3}, \frac{2}{3})$. To obtain A_2, remove from A_1 its middle thirds, namely $(\frac{1}{9}, \frac{2}{9})$ and $(\frac{7}{9}, \frac{8}{9})$. Then set

$$A_n = A_{n-1} \setminus \bigcup_{p=0}^{\infty} \left(\frac{1+3p}{3^n}, \frac{2+3p}{3^n} \right).$$

DEFINITION. *The **Cantor set**, denoted by C, is the intersection (over $n \in \mathbb{N}$) of the A_n introduced just above, that is:*

$$C = \bigcap_{n \in \mathbb{N}} A_n.$$

REMARK. The Cantor set is uncountable as will be seen in later chapters. Its topological properties will be considered throughout the present manuscript.

In measure theory, the Cantor set constitutes an example of an uncountable set with zero Lebesgue measure.

1.2. Exercises With Solutions

Exercise 1.2.1. Let I be an arbitrary set. Let $f : X \to Y$ and $g : Y \to Z$ be two functions. Also assume that A, $A_i \subset X$ and B, $B_i \subset Y$. Show that the following statements hold

(1) $f(\bigcup_{i \in I} A_i) = \bigcup_{i \in I} f(A_i)$;

(2) $f(\bigcap_{i \in I} A_i) \subseteq \bigcap_{i \in I} f(A_i)$;

(3) $f^{-1}(\bigcup_{i \in I} B_i) = \bigcup_{i \in I} f^{-1}(B_i)$;

(4) $f^{-1}(\bigcap_{i \in I} B_i) = \bigcap_{i \in I} f^{-1}(B_i)$;

(5) $f^{-1}(B^c) = [f^{-1}(B)]^c$;

(6) $(g \circ f)^{-1}(A) = f^{-1}(g^{-1}(A))$;

(7) $f(f^{-1}(A)) \subset A$;

(8) $B \subset f^{-1}(f(B))$.

Exercise 1.2.2. Let $f : X \to Y$ be a function. If f_A is the restriction of f to $A \subset X$, then show that for any subset U of X one

has
$$f_A^{-1}(U) = A \cap f^{-1}(U).$$

Exercise 1.2.3.

(1) Show, using the map
$$f : \mathbb{N} \times \mathbb{N} \to \mathbb{N}$$
$$(n, m) \mapsto f(n, m) = 2^n 3^m,$$
that $\mathbb{N} \times \mathbb{N}$ is countable.

(2) Deduce that if A and B are two countable sets, then so is their Cartesian product $A \times B$.

Exercise 1.2.4. Let $\mathbb{Q}[X]$ be the set of polynomials with *rational* coefficients. Show that $\mathbb{Q}[X]$ is countable.

Exercise 1.2.5. Let $X = \{0, 1\}^{\mathbb{N}}$ be the set of all sequences having values in $\{0, 1\}$. Show that X is uncountable.

Exercise 1.2.6. Show that

(1) $\displaystyle\bigcap_{n \in \mathbb{N}} \left(\frac{-1}{n}, \frac{1}{n} \right) = \{0\}$;

(2) $\displaystyle\bigcup_{n \in \mathbb{N}} [-n, n] = \mathbb{R}$;

(3) $\displaystyle\bigcap_{n \in \mathbb{N}} [n, \infty) = \varnothing$.

Exercise 1.2.7. Let $n \geq 1$. Find $\displaystyle\bigcap_n A_n$ and $\displaystyle\bigcup_n A_n$ in the following cases

(1) $A_n = \{1, 2, \cdots, n\}$;
(2) $A_n = (-n, n)$;
(3) $A_n = \left(-\frac{1}{n}, 1 + \frac{1}{n} \right)$;
(4) $A_n = \left[0, 1 - \frac{1}{n} \right]$;
(5) $A_n = \left(-\frac{1}{n}, 1 \right)$.

Exercise 1.2.8. [Young's Inequality] Let a and b be two positive real numbers. Let $p > 1$ and $q > 1$ be such that $\frac{1}{p} + \frac{1}{q} = 1$ (q is called the conjugate of p). Show that we have
$$ab \leq \frac{a^p}{p} + \frac{b^q}{q}.$$

Exercise 1.2.9. (A bit of number theory!) Show that $e \notin \mathbb{Q}$ (you may use the sequences defined as $x_n = \displaystyle\sum_{k=0}^{k=n} \frac{1}{k!}$ and $y_n = x_n + \frac{1}{n!n}$).

Exercise 1.2.10. Show that the set of rational numbers $\mathbb{Q}$ is dense in $\mathbb{R}$, that is, between any two real numbers, there is a rational one.

REMARK. The precise meaning of density will be made clear in its general context in Chapter 3.

Exercise 1.2.11. Let $X = C([0,1])$ and let $f, g \in X$. Show that

$$\int_0^1 |f(x)g(x)|dx \leq \sup_{0 \leq x \leq 1} |f(x)| \int_0^1 |g(x)|dx$$

REMARK. The main purpose of the foregoing exercise is to familiarize the reader (especially the students) with the result therein. Usually, when it is given like that they think that we have just taken $|f(x)|$ outside the integral which is of course not true. In the solution below we show how to deal with this situation. This is something that must be remembered. A similar idea will be applied to infinite series in due time.

1.3. More Exercises

Exercise 1.3.1. Let X and Y be two sets where $A \subset X$. Let $f : X \to Y$ be a function. Do we have

$$f(A^c) = (f(A))^c?$$

Exercise 1.3.2. Let X and Y be two sets. Let $f : X \to Y$ be a function. Show that:

(1) f is injective iff $f(A \cap B) = f(A) \cap f(B)$ for all $A, B \subset X$ iff $f^{-1}(f(C)) = C$ for all $C \subset X$.
(2) f is surjective iff $f(f^{-1}(D)) = D$ for all $D \subset Y$.

Exercise 1.3.3 (π irrational)**.** Let P be a polynomial of degree $2n$. Set

$$F(x) = P(x) - P''(x) + P^{(4)}(x) - \cdots + (-1)^n P^{(2n)}(x).$$

(1) By observing that $P(x)\sin x = (F'(x)\sin x - F(x)\cos x)'$, show that

$$\int_0^\pi P(x)\sin x\, dx = F(0) + F(\pi) \text{ (called Hermite's Formula).}$$

(2) Assume that π is rational, i.e. $\pi = \frac{a}{b}$ with $a \in \mathbb{N}$ and $b \in \mathbb{N}$.
  (a) Apply Hermite's formula to $P(x) = \frac{1}{n!}x^n(a - bx)^n$.
  (b) Show that $P(0) = P'(0) = \cdots P^{(n-1)}(0) = 0$ and deduce that $F(0), F(\pi) \in \mathbb{Z}$.

(c) Now set $I_n = \int_0^\pi P(x) \sin x \, dx$. Prove that $I_n > 0$ and that $\lim\limits_{n \to +\infty} I_n = 0$. Deduce from Question b) that $I_n \in \mathbb{Z}$ and find a contradiction (leading to the irrationality of π).

Exercise 1.3.4. Show that between any two (different) reals, there is an irrational.

Exercise 1.3.5. Show that $\{q\sqrt{3} : q \in \mathbb{Q}\}$ is dense in $\mathbb{R} \setminus \mathbb{Q}$.

Exercise 1.3.6. Exhibit an explicit dense subset of $\mathbb{C}$. The same question for $\mathbb{C}^n$.

Exercise 1.3.7. Show that the direct image of a countable set *under any function* remains countable.

Exercise 1.3.8.

(1) Show that the countable union of countable sets is countable.
(2) Deduce that $\mathbb{Q}$ is countable too.

Exercise 1.3.9. An **algebraic number** is a root of a polynomial having rational (or integer) coefficients. The set of algebraic numbers certainly includes $\mathbb{Q}$. It also includes many irrational numbers such as $\sqrt{2}$, $\sqrt{3}$, $\frac{1}{\sqrt{2}}$ and $1 + \sqrt[3]{5}$...etc.

A non-algebraic number is called **transcendental**. In number theory, it is known that π or e are transcendental.

(1) Show that the set of algebraic numbers is countable.
(2) Need the set of transcendental numbers be countable? Why?
(3) Show that the sets of algebraic and transcendental numbers are both dense in $\mathbb{R}$.

Exercise 1.3.10. Let $A \subset \mathbb{R}$ be non void and bounded. Show that

$$\sup_{x,y \in A} |x - y| = \sup A - \inf A.$$

CHAPTER 2

Metric Spaces

2.1. What You Need to Know

2.1.1. Definitions and Examples.

DEFINITION. *Let X be a non-empty set. A **metric** (or a **distance**) on X is a function $d : X \times X \to \mathbb{R}^+$ verifying:*

(1) $d(x, y) = 0 \iff x = y$.

(2) $\forall x, y \in X : \ d(x, y) = d(y, x)$.

(3) $\forall x, y, z \in X : \ d(x, z) \le d(x, y) + d(y, z)$ *(Triangle Inequality).*

*The couple (X, d) is called a **metric space**.*

EXAMPLES.

(1) *The mapping $d : \mathbb{R} \times \mathbb{R} \to \mathbb{R}^+$ defined by $d(x, y) = |x - y|$ is a metric called the **usual** (or **standard**) metric on $\mathbb{R}$. It may be denoted by $| \cdot |$.*

(2) *The mapping $d : \mathbb{C} \times \mathbb{C} \to \mathbb{R}^+$ defined by $d(z, z') = |z - z'|$ is a metric called the **usual** (or **standard**) metric on $\mathbb{C}$. It may also be denoted by $| \cdot |$ or $| \cdot |_{\mathbb{C}}$.*

Here is another example.

EXAMPLE (the discrete metric). *Let X be a non-empty set. Define a map on $X \times X$ by*

$$d(x, y) = \left\{ \begin{array}{ll} 0, & x = y, \\ 1, & x \neq y. \end{array} \right.$$

*Then d is a metric on X called the **discrete metric** (for a proof and some properties of d, see Exercise 2.3.5).*

REMARK. The discrete metric is not very useful in practise since it can be defined on any non-empty set. However, it can be very valuable as a source for counterexamples as will be illustrated in many exercises in the sequel.

2.1.2. Important Sets in Metric Spaces.

DEFINITION. *Let (X, d) be a metric space.*

An **open ball** of center $a \in X$ and radius $r > 0$, denoted by $B(a,r)$, is defined by:

$$B(a,r) = \{x \in X : \ d(x,a) < r\}.$$

A **closed ball** of center $a \in X$ and radius $r > 0$, denoted by $B_c(a,r)$, is defined by:

$$B_c(a,r) = \{x \in X : \ d(x,a) \leq r\}.$$

A **sphere** of center $a \in X$ and radius $r > 0$, and denoted by $S(a,r)$, is given by:

$$S(a,r) = \{x \in X : \ d(x,a) = r\}.$$

DEFINITION. Let (X,d) be a metric space and let $U \subset X$. Then we say that U is **open** in (X,d) (or just in X) if:

$$\forall x \in U, \exists r > 0 : \ B(x,r) \subset U.$$

A subset V of X is said to be **closed** if its algebraic complement V^c is open.

EXAMPLES.

(1) An open interval is an open set (see Exercise 2.3.17 for a more general result).
(2) $[0,1]$ or $[0,1)$ are not open in usual $\mathbb{R}$.
(3) In usual $\mathbb{R}$ again, $\{0\}$ is not open.

REMARK. Open and closed sets do not form a partition of the metric space. There are sets which are neither open nor closed. The reader will see many examples throughout the exercises in Chapters 2 & 3.

There are also sets which are open and closed simultaneously (for example in a discrete metric space, see Exercise 2.3.5). We call them **"clopen"**.

REMARK. It is clear that we can define different metrics on the same set X. Hence a given subset A of X may be open with respect to a metric and not open with respect to another one.

THEOREM (See Exercise 2.3.15). Let (X,d) be a metric space. Then

(1) $\varnothing$ and X are open.
(2) The union of **any** collection of open sets remains open.
(3) The intersection of a **finite** collection of open sets remains open.

REMARK. We will see in the next chapter that a space in which the properties of the previous theorem are verified will be called a **topological space**. Thus a metric space is an example (very important though) of a topological space.

EXAMPLES.
(1) $[0, 1]$ *is closed in usual* $\mathbb{R}$ *for its complement* $(-\infty, 0) \cup (1, \infty)$, *being a union of open sets, is open.*
(2) *A similar idea gives us the closedness of* $\{0\}$ *in usual* $\mathbb{R}$.

COROLLARY (See Exercise 2.3.16). *A subspace of a metric space is open iff it is a union of open balls.*

EXAMPLE. *Open sets in usual* $\mathbb{R}$ *are unions of open intervals.*

The definition of closed sets combined with elementary set theory yield

COROLLARY. *Let* (X, d) *be a metric space. Then*
(1) $\varnothing$ *and* X *are closed.*
(2) *The intersection of* **any** *collection of closed sets stays closed.*
(3) *The union of a* **finite** *collection of closed sets remains closed.*

Now we introduce the notion of a bounded set.

DEFINITION. *Let* (X, d) *be a metric space and let* $A \subset X$. *Define:*
$$d(A) = \sup_{x,y \in A} d(x, y).$$
Then $d(A)$ *is called the* **diameter** *of* A.

REMARK. If A is a non-empty set, then $d(A) \leq \infty$.

DEFINITION. *Let* (X, d) *be a metric space and let* $A \subset X$. *We say that* A *is* **bounded** *if it is contained in a ball of a* **finite** *radius. Equivalently,* A *is bounded if its diameter* $d(A)$ *is finite.*

EXAMPLE. *Usual* $\mathbb{R}$ *is not bounded while* $\mathbb{R}$ *equipped with a discrete metric is bounded.*

DEFINITION. *Let* X *be a non-empty set. Let* (Y, d) *be a metric space. A function* $f : X \to (Y, d)$ *is said to be* **bounded** *if* $f(X)$ *is bounded in* (Y, d).

2.1.3. Continuity in Metric Spaces.

DEFINITION. *Let* (X, d) *and* (Y, d') *be two metric spaces. Let* $f : (X, d) \to (Y, d')$ *be a function. Then* f *is* **continuous** *at a point* $a \in X$ *if for every ball* $B(f(x), \varepsilon)$, *there exists a ball* $B(x, d')$ *such that* $f(B(x, d')) \subset B(f(x), \varepsilon)$.

This is the natural extension of the known definition of a continuous real-valued function defined on an interval of $\mathbb{R}$. It is equivalent to the following result (which will be the actual definition of a continuous function between two topological spaces):

PROPOSITION (See Exercise 2.3.21). *Let (X, d) and (Y, d') be two metric spaces. Let $f : (X, d) \to (Y, d')$ be a function. Then f is continuous iff for every open set U in (Y, d'), $f^{-1}(U)$ is open in (X, d).*

DEFINITION. *Let (X, d) be a metric space and let $A \subset X$ be non-empty. The distance between a point $x \in X$ and A, denoted by $d(x, A)$ is defined by:*

$$d(x, A) = \inf_{t \in A} d(x, t).$$

REMARK. The function $x \mapsto d(x, A)$ is continuous on X thanks to the following result:

$$\forall x, y \in X : \ |d(x, A) - d(y, A)| \leq d(x, y).$$

We can also define uniform continuity of a function f between two metric spaces.

DEFINITION. *Let (X, d) and (X, d') be two metric spaces and let $f : (X, d) \to (X, d')$ be a function. We say that f is **uniformly continuous** if:*

$$\forall \varepsilon > 0, \exists \alpha > 0, \ \forall x, x' \in X : \ (d(x, x') < \alpha \implies d'(f(x), f(x')) < \varepsilon).$$

REMARK. It is plain that uniform continuity implies continuity.

DEFINITION. *Let (X, d) and (Y, d') be two metric spaces. A function $f(X, d) \to (Y, d')$ is said to be an **isometry** if f is surjective and:*

$$\forall x, y \in X : \ d(x, y) = d'(f(x), f(y)).$$

REMARKS.

(1) In some textbooks, they do not assume the surjectivity hypothesis in the definition of an isometry (they then say **an isometry into**).
(2) It is plain that an isometry is injective. Thus an isometry is bijective.
(3) The inverse of an isometry is an isometry.
(4) It is also clear that an isometry is continuous.

2.1.4. Equivalent Metrics.

DEFINITION. *Let d and d' be two metrics on the same set X. Then we say that d is **equivalent** (we may also say **strongly equivalent** or **Lipschitz equivalent**) to d' if:*

$$\exists \alpha, \beta > 0, \ \forall x, y \in X \ : \ \alpha d(x,y) \le d'(x,y) \le \beta d(x,y).$$

There is another type of equivalence of metrics, namely

DEFINITION. *Two metrics d and d' on a set X are said to be **topologically equivalent** if: a subset U is open in (X, d) iff it is open in (X, d').*

REMARK. (Strong) equivalence of two metrics implies topological equivalence, but not vice versa. For a counterexample, see e.g. Exercise 2.3.26.

EXAMPLES.

(1) *The metrics d_p with $1 \le p \le \infty$, defined in Exercise 2.3.9, are equivalent, hence topologically equivalent.*

(2) *In $\mathbb{R}$, the discrete metric is not topologically equivalent to the usual metric for an open set w.r.t. the discrete metric need not be open w.r.t. the usual one. This also implies that the two metrics are not (strongly) equivalent.*

2.2. True or False: Questions

QUESTIONS. Comment on the following questions/statements and indicate those which are false and those which are true when this applies. Justify your answers.

(1) Let X be set with $\operatorname{card} X \ge 2$. We can always define a metric on X.

(2) Let d be a metric on some set X. Then d is a positive function.

(3) The set $\{0\}$ is not open.

(4) The set $\{0\}$ is not open in $(\mathbb{R}, |\cdot|)$ since it is closed.

(5) In a metric space, a ball can contain another ball of strictly bigger radius.

(6) The set $(0, 1)$ is bounded.

(7) Let (X, d) be a metric space and let A, B and C be three subsets of X. Define (not to be confused with the diameter of a given set)

$$d(A, B) = \inf_{(a,b) \in A \times B} d(a, b).$$

Then

$$d(A, C) \le d(A, B) + d(B, C).$$

(8) Let (X, d) and (X, d') be two metric spaces such that d and d' are equivalent. Show that

U open in (X, d) if and only if U open in (X, d').

(9) Every closed ball is a closed set. What about the converse?
(10) The sphere is a closed set.
(11) There is a metric space in which all triangles are isosceles.

2.3. Exercises With Solutions

Exercise 2.3.1. Assume that a function d on $X \times X$ into $\mathbb{R}^+$ verifies

$$\begin{cases} d(x, y) = 0 \Leftrightarrow x = y, \\ d(x, y) = d(y, x), & \forall x, y \in X, \\ d(x, z) \leq d(x, y) + d(y, z), & \forall x, y, z \text{ distinct and in } X. \end{cases}$$

Show that d is a metric on X.

Exercise 2.3.2. Let (X, d) be a metric space. Show that

$$\forall x, y, z \in X : \ |d(x, z) - d(y, z)| \leq d(x, y).$$

Exercise 2.3.3. Are the following functions metrics on X?

(1) $d(x, y) = |x^2 - y^2|$, $X = \mathbb{R}$;
(2) $d(x, y) = |x^3 - y^3|$, $X = \mathbb{R}$;
(3) $d(x, y) = e^{x-y}$, $X = \mathbb{R}$;
(4) $d(x, y) = \left| \frac{1}{x} - \frac{1}{y} \right|$, $X = \mathbb{R}^*$;
(5) $d(x, y) = |x - 3y|$, $X = \mathbb{R}$.

Exercise 2.3.4. In the usual metric of $\mathbb{R}$, what is the ball corresponding to the open interval $(0, 1)$ (i.e. what is its center and what is its radius?).

Exercise 2.3.5 (the discrete metric). Let X be a non-empty set. Define a map on $X \times X$ by

$$d(x, y) = \begin{cases} 0, & x = y, \\ 1, & x \neq y. \end{cases}$$

(1) Show that d is a metric on X.
(2) Let $r > 0$ and let $x \in X$. Find the open ball $B(x, r)$ and the closed ball $B_c(x, r)$.
(3) Find the sphere $S(x, r)$.
(4) Show that every subset in a discrete metric space is open.
(5) Deduce that every subset in a discrete metric space is closed.
(6) Set $X = \mathbb{R}$. Show that the discrete metric on $\mathbb{R}$ is not equivalent to the usual metric on $\mathbb{R}$.

Exercise 2.3.6. Let (X, d) be a metric space. Show that $d'(x, y) = \sqrt{d(x, y)}$ defines a metric on X.

Exercise 2.3.7. On $\mathbb{N} \times \mathbb{N}$, we define

$$d(x, y) = \begin{cases} 0, & x = y, \\ 3 + \frac{x+y}{xy}, & x \neq y. \end{cases}$$

Show that d is a metric on $\mathbb{N}$.

Exercise 2.3.8. On $\mathbb{R}^n \times \mathbb{R}^n$, define the function d by

$$d(x, y) = \sum_{k=1}^{n} |x_k - y_k|$$

for all $x = (x_k), y = (y_k) \in \mathbb{R}^n$. Show that d is a metric on $\mathbb{R}^n$ (called in many references the *taxicab* metric).

Exercise 2.3.9. Let $p > 1$ and $q > 1$ be such that $\frac{1}{p} + \frac{1}{q} = 1$. Let $a_1, \cdots, a_n$ and $b_1, \cdots, b_n$ be positive real numbers.

(1) Prove the following Hölder's inequality

$$\sum_{k=1}^{n} a_k b_k \leq \left(\sum_{k=1}^{n} (a_k)^p \right)^{\frac{1}{p}} \left(\sum_{k=1}^{n} (b_k)^q \right)^{\frac{1}{q}}.$$

(2) Prove the following so-called Minkowski's inequality

$$\left(\sum_{k=1}^{n} (a_k + b_k)^p \right)^{\frac{1}{p}} \leq \left(\sum_{k=1}^{n} (a_k)^p \right)^{\frac{1}{p}} + \left(\sum_{k=1}^{n} (b_k)^p \right)^{\frac{1}{p}}.$$

(3) Let $p \geq 1$. Define the following functions on $\mathbb{R}^n \times \mathbb{R}^n$

$$d_p(x, y) = \left(\sum_{k=1}^{n} |x_k - y_k|^p \right)^{\frac{1}{p}} \quad \text{and} \quad d_\infty(x, y) = \max_{1 \leq k \leq n} (|x_k - y_k|).$$

(a) Show that d_p and d_∞ are metrics on $\mathbb{R}^n$.
(b) Show that they are equivalent metrics and deduce that

$$\lim_{p \to \infty} d_p(x, y) = d_\infty(x, y).$$

REMARKS.

(1) If $p = 1$, then we get back the taxicab metric whose proof does not require the Minkowski inequality.
(2) If $p = 1$, then in general, we allow q to be ∞ and the same applies for $q = 1$. This is, however, not discussed in this exercise.

(3) If $p = q = 2$, we get back one of the versions of the Cauchy-Schwarz inequality . Therefore, Hölder's Inequality generalizes Cauchy-Schwarz's.

(4) From Question 3, b), we now know why we use the notation d_∞.

Exercise 2.3.10. Let $M = (X, d)$ be a metric space.

(1) Show that both

    (a) $\delta(x, y) = \min(1, d(x, y))$ and

    (b) $\rho(x, y) = \frac{d(x,y)}{1+d(x,y)}$

    define metrics on X (hint: for ρ you may start by showing that $0 \leq x \leq y \Rightarrow \frac{x}{1+x} \leq \frac{y}{1+y}$ and $\frac{x+y}{1+x+y} \leq \frac{x}{1+x} + \frac{y}{1+y}$, $\forall x, y \geq 0$).

(2) Regardless of what X can be, is (X, δ) bounded? Is (X, ρ) bounded?

Exercise 2.3.11. Let (X_n, d_n), $n = 1, 2, \cdots$ be a countable family of metric spaces. Set $X = \prod_{n=1}^{\infty} X_n$. Show that the function $d : X \times X \to \mathbb{R}$ defined by,

$$d(x, y) = \sum_{n=1}^{\infty} \frac{1}{2^n} \times \frac{d_n(x_n, y_n)}{1 + d_n(x_n, y_n)},$$

for each $x = (x_1, \cdots, x_n, \cdots)$ et $y = (y_1, \cdots, y_n, \cdots)$ in X, is a metric on X.

Exercise 2.3.12. Let X be the space of real-valued continuous functions on $[0, 1]$.

(1) Show that

$$d(f, g) = \int_0^1 |f(x) - g(x)|dx \text{ and } d'(f, g) = \sup_{x \in [0,1]} |f(x) - g(x)|$$

are two metrics on X (d' is usually called the **supremum metric**).

(2) Show that these two metrics are not equivalent.

(3) Does d remain a metric if X is replaced by the space of Riemann-integrable functions?

Exercise 2.3.13. Let $X = C^1([0, 1], \mathbb{R})$ be the space of real-valued continuous functions defined, differentiable and having a continuous derivative on $[0, 1]$. Define a function from $X \times X$ into $\mathbb{R}^+$ by

$$d(f, g) = \sup_{x \in [0,1]} |f'(x) - g'(x)|$$

where f' stands for the derivative of f. Is d a metric on X?

Exercise 2.3.14. Let $M = (X, d)$ be a metric space. Also, let $x, y \in X$ and $r, s > 0$.

(1) Show that $B(x, r) = B(y, s)$ (for all x, y and all r, s) does not always give $x = y$ or $r = s$.

(2) Give one case when this is always correct.

Exercise 2.3.15. Let (X, d) be a metric space. Show that:

(1) X and $\varnothing$ are open sets in (X, d),

(2) The arbitrary union of open sets in (X, d) is open in (X, d),

(3) The finite union of open sets in (X, d) is open in (X, d). Does this stays true for an infinite union?

Exercise 2.3.16. Show that in any metric space X, a subset U of X is open if and only if it can be written as a union of open balls.

Exercise 2.3.17. Let (X, d) be a metric space. Show that an open ball is an open set. Is the converse always true?

Exercise 2.3.18. Are the intervals $[a, b]$, $[a, b)$ or $(a, b]$ open in $\mathbb{R}$?

Exercise 2.3.19. Associate with $\mathbb{R}^2$ the euclidian metric and denote it by d. Define on $\mathbb{R}^2 \times \mathbb{R}^2$ a function δ by

$$\delta(x, y) = \begin{cases} 0, & x = y, \\ d(x, \mathbf{0}) + d(y, \mathbf{0}), & x \neq y \end{cases}$$

where $\mathbf{0} = (0, 0)$.

(1) Check that δ is in effect a metric on $\mathbb{R}^2$.

(2) Let $a \neq \mathbf{0}$. Show that $\{a\}$ is open.

(3) Is $\{\mathbf{0}\}$ open in $(\mathbb{R}^2, \delta)$?

(4) What is $\mathbb{R}^2 \setminus \{\mathbf{0}\}$ restricted to δ?

REMARK. The metric defined in the previous exercise has a particular name: In the UK, it is called the *British Rail Metric*. The main reason for this designation is that often when one wants to travel from a town to another, then he/she might have to pass by some train station in London which is represented by $\mathbf{0}$ in our exercise. For the same reason, the French call it the *SNCF Metric*. As for the Americans, they call it the *Post Office Metric*. This latter is probably more meaningful than the other two as if one wants to send a letter from his/her place to somewhere else it will have to pass by the post office which is represented by $\mathbf{0}$ in the exercise.

Exercise 2.3.20 (Ultrametric space). Let X be a non empty set. Let d be a function from $X \times X$ into $\mathbb{R}^+$ such that

- $d(x, y) = 0 \Leftrightarrow x = y$,

- $d(x, y) = d(y, x), \forall x, y \in X,$
- $d(x, z) \leq \max(d(x, y), d(y, z)), \forall x, y, z \in X.$

(1) Show that d is a metric, called **ultrametric**.
(2) Show that at least two of $d(x, y)$, $d(y, z)$ and $d(x, z)$ must be equal. Interpret this result geometrically.
(3) Show that in a ultrametric space, every point in an open ball is its center.
(4) Show that open balls are clopen and that so are closed balls too.

Exercise 2.3.21. Let $f : X \to X'$ be a function where X and X' are two topological spaces. Show that f is continuous on X iff whenever U is an open set in X', $f^{-1}(U)$ is an open set in X.

Exercise 2.3.22. Let (X, d) be a metric space. Let
(1) Show that the function $f : X \to \mathbb{R}$ defined for all $x \in X$ by $f(x) = d(x, a)$ is continuous on X where $a \in X$ and $r > 0$.
(2) Let $B \subset X$ be non-empty. Set

$$g(x) = d(x, B) = \inf_{b \in B} d(x, b).$$

Show that g is uniformly continuous on X.

Exercise 2.3.23. Let $\mathbb{R}^+$ be equipped with the induced usual metric $|\cdot|$. Let d be the *metric*, defined for all $x, y \in \mathbb{R}^+$, by

$$d(x, y) = |\sqrt{x} - \sqrt{y}|.$$

Let $f : (\mathbb{R}^+, |\cdot|) \to (\mathbb{R}^+, d)$ be the identity map.
(1) Show that f is uniformly continuous.
(2) Interpret the result of the previous question differently.

Exercise 2.3.24. In usual $\mathbb{R}$, give an example of a:
(1) function $f : \mathbb{R} \to \mathbb{R}$ not continuous at every point,
(2) function $f : \mathbb{R} \to \mathbb{R}$ continuous at only one point,
(3) function $f : \mathbb{R} \to \mathbb{R}$ continuous at only two points.

Exercise 2.3.25. Show that the open ball $B(a, r)$ is open in the metric space (X, d) by using the function $f : x \mapsto f(x) = d(x, a)$ where $a \in X$.

Exercise 2.3.26. Let x and y be two reals. Set for all x and y

$$\delta(x, y) = |\arctan x - \arctan y|.$$

(1) Show that δ is a metric on $\mathbb{R}$. Is $\mathbb{R}$ bounded with respect to this metric?

(2) Compare the open balls $B(0,2)$, $B(0,4)$ and $B(1,4)$. What do you observe?

(3) Show that δ is not equivalent to the usual metric on $\mathbb{R}$.

(4) Is δ topologically equivalent to the usual metric on $\mathbb{R}$?

Exercise 2.3.27. Let (X,d) be a metric space. Let ρ be as in Exercise 2.3.10.

(1) Is d equivalent to ρ?

(2) Are they topologically equivalent?

2.4. Tests

Test 1. Show that even if a metric d is not defined into $\mathbb{R}^+$, the three properties of a metric will guarantee its positivity, i.e.

$$d(x,y) \geq 0 \text{ for all } x,y \in X.$$

Test 2. Let $X = \mathbb{R}$. We define on $\mathbb{R} \times \mathbb{R}$, the function d by

$$d(x,y) = \ln(1 + |x - y|), \ \forall x,y \in \mathbb{R}.$$

Show that d is a metric on X.

Test 3. Let $X = C([0,1], \mathbb{R})$. Define a function d on $X \times X$ by

$$d(f,g) = \left(\int_0^1 |f(x) - g(x)|^2 dx \right)^{\frac{1}{2}}$$

where $f, g \in X$. Show that d is a metric on X.

Test 4. Let $X = \{x, y\}$. Let $a > 0$. Assume that a function d on $X \times X$ verifies

$$\begin{cases} d(x,x) = d(y,y) = 0, \\ d(x,y) = d(y,x) = a. \end{cases}$$

Show that d is a metric on X.

Test 5. Let (X,d) be a metric space and let A be some non-empty set. Let $f : A \to X$ be a *one-to-one* mapping. Set

$$\delta(x,y) = d(f(x), f(y)), \ \forall x,y \in A.$$

Show that δ defines a metric on A.

Test 6. Give an example of a metric which is ultrametric and an example of one which is not.

2.5. More Exercises

Exercise 2.5.1. Let (X, d) be a metric space. Show that

$$d(x_1, x_n) \leq d(x_1, x_2) + d(x_2, x_3) + \cdots + d(x_{n-1}, x_n), \ \forall x_1, x_2, \cdots, x_n \in X.$$

Exercise 2.5.2. Let (X, d) be a metric space.

(1) Show that

$$|d(x, z) - d(y, t)| \leq d(x, y) + d(z, t).$$

(2) Why can this result be considered as a generalization of the inequality appearing in Exercise 2.3.2?

Exercise 2.5.3. Let X be a non-void set. Assume that a function d on $X \times X$ verifies

$$\begin{cases} d(x, x) = 0, \\ 1 \leq d(x, y) = d(y, x) \leq 2, \quad x \neq y. \end{cases}$$

Show that d is a metric on X.

Exercise 2.5.4. Let d be a function defined on $\mathbb{C} \times \mathbb{C}$ by

$$d(z, z') = \begin{cases} 0, & z = z', \\ |z| + |z'|, & z \neq z'. \end{cases}$$

(1) Prove that d is a metric on $\mathbb{C}$.
(2) Is d topologically equivalent to the usual metric on $\mathbb{C}$?

Exercise 2.5.5. Describe the open sets in the metric spaces of Exercise 2.3.19.

Exercise 2.5.6. Let (X, d) be a metric space.

(1) Show that the finite union of bounded sets is bounded.
(2) What about the arbitrary union?
(3) Show that the arbitrary intersection of bounded sets is bounded.

Exercise 2.5.7. Is d a metric on X in the following cases:

(1) $d(f, g) = \sup_{x \in [0,1]} |f'(x) - g'(x)| + |f(0) - g(0)|$ where f' stands for the derivative of f, $X = C^1([0, 1], \mathbb{R})$ (cf Exercise 2.3.13).
(2) $d(f, g) = \sup_{x \in [a,b]} |(f(x) - g(x))\omega(x)|$ where $\omega \in X$ (a **weight**) be not vanishing on $[a, b]$, $X = C([a, b])$.
(3) $d(f, g) = \sqrt[p]{\int_0^1 |f(x) - g(x)|^p dx}$ where $p \geq 1$, $X = C([0, 1], \mathbb{R})$ (hint: use Exercise 2.3.9).

Exercise 2.5.8. Assume that (X, d_k) are metric spaces for all $1 \leq k \leq n$. Let $p \geq 1$. Show that

$$d'(x, y) = \left(\sum_{k=1}^{n} d_k(x_k, y_k)^p \right)^{\frac{1}{p}}$$

defines a metric on X^n. Does d remain a metric if $0 \leq p < 1$?

Exercise 2.5.9. Let $A = (0, 2) \times \{1\}$. Show that A is open in $\mathbb{R} \times \{1\}$. Is A open in $\mathbb{R}^2$?

Exercise 2.5.10. Let X be a non-empty set. Define a map on $X \times X$ by

$$\delta(x, y) = \begin{cases} 0, & \text{if } x = y, \\ 2, & \text{if } x \neq y. \end{cases}$$

(1) Show that δ is a metric on X.
(2) Is δ equivalent to the discrete metric? Are they topologically equivalent?

Exercise 2.5.11. On $\mathbb{N} \times \mathbb{N}$, define a function d by $d(x, x) = 0$ for every $x \in \mathbb{N}$ and for $x, y \in \mathbb{N}$, $x \neq y$ by

$$d(x, y) = 5 + \frac{1}{x} + \frac{1}{y}.$$

Show that d is a metric on $\mathbb{N}$.

Exercise 2.5.12. Show that the metrics defined in Exercise 2.3.12 are not topologically equivalent.

Exercise 2.5.13. For all $x, y > 0$, let

$$d(x, y) = |\ln x - \ln y|.$$

(1) Check that d is a metric on $\mathbb{R}_+^*$.
(2) Prove that d is topologically equivalent to the induced usual metric on $\mathbb{R}_+^*$.

CHAPTER 3

Topological Spaces

3.1. What You Need to Know

3.1.1. General Notions.

DEFINITION. *Let X be a non-empty set. A **topology** T on X is a subset of $\mathcal{P}(X)$ verifying the following axioms:*

(1) *$\varnothing$,X are both in T.*

(2) *The intersection of two (hence of a **finite** collection of) sets in T remains in T.*

(3) *The **arbitrary** union of sets in T is again in T.*

*The couple (X, T) is then called a **topological space**.*
*Elements of T are called **open**.*

EXAMPLES.

(1) *Let X be a set. Then $T = \{\varnothing, X\}$ is a topology on X called the **indiscrete** (or **trivial**) topology. Notice that this topology is not too interesting as there are practically no open or closed sets.*

(2) *Also, take $T = \mathcal{P}(X)$, i.e. the collection of all subsets of X. Then T is a topology on X. It is called the **discrete topology**.*

(3) *Every metric metric space is a topological space. Hence, the usual metric on $\mathbb{R}$ gives us a topology which we call the **usual topology**.*

(4) *Let $\{0, 1, 2\}$ and let*
$$T = \{\varnothing, \{0\}, \{2\}, X\}.$$
Then T is not a topology on X.

DEFINITION. *Let (X, T) be a topological space. A set $V \subset X$ is said to be **closed** if V^c is open, that is, if $V^c \in T$.*

EXAMPLES.

(1) *In standard $\mathbb{R}$, $\mathbb{Z}$ is closed for its complement is a (here an infinite) union of open intervals, hence $\mathbb{Z}^c$ is an open set in $\mathbb{R}$.*

(2) *In a metric space, we have seen that any closed ball is a closed set.*

REMARK. As in the case of metric spaces, there are:

(1) Open sets which are not closed.
(2) Closed sets which are not open.
(3) Sets which are neither open nor closed.
(4) Sets which are open and closed at the same time (we call them "**clopen**").

By elementary set theory, we have:

PROPOSITION. *Let (X, T) be a topological space. Then*

(1) *$\varnothing$ and X are closed.*
(2) *The union of a **finite** collection of closed sets is closed.*
(3) *The **arbitrary** intersection of closed sets is closed.*

We saw above that every metric space is a topological space. We may therefore ask whether any topological space can be seen as a metric space in some sense? Giving this a specific terminology seems to be appropriate. We have

DEFINITION. *Let (X, T) be a topological space. We say that T is **metrizable** if there is some metric d (on X) which gives the topology of T.*

EXAMPLES.

(1) *A discrete topological space is metrizable. We can easily verify that the discrete metric induces the discrete topology.*
(2) *An indiscrete topological space X (with $\operatorname{card} X \geq 2$) is not metrizable.*

REMARK. There are theorems giving conditions under which a given topological space is metrizable, such as the "Urysohn Metrization Theorem". But this is not within the scope of the present book. For more details, see e.g. [**10**].

DEFINITION. *Let (X, T) be a topological space and let $x \in X$. A **neighborhood** of x is any **open** set $U \in T$ which contains x.*
The set of neighborhoods of x is denoted by $\mathcal{V}(x)$.

REMARK. In some textbooks, neighborhoods are not taken to be open. For further discussion, see the "True or False" section.

DEFINITION. *Let T and T' be two topologies on a set X. We say that:*

(1) *T is finer than T' if $T' \subset T$;*
(2) *T is coarser than T' if $T' \supset T$.*

*If T and T' are such that $T' \subset T$ **or** $T \subset T'$, then we say that T and T' are **comparable**.*

Examples.

(1) *The indiscrete topology is coarser than any other topology which may be defined on the same set.*
(2) *The discrete topology is finer than any other topology which may be defined on the same set.*

Definition. *A topological space (X, T) is called **Hausdorff** (or **separated**) if:*

$$\forall x, y \in X, \ x \neq y, \ \exists (U, V) \in \mathcal{V}(x) \times \mathcal{V}(y) : \ U \cap V = \varnothing.$$

Examples.

(1) *Any metric space is Hausdorff (Exercise 3.5.10).*
(2) *An indiscrete topological space is not Hausdorff.*

Proposition (See Exercise 3.3.28). *Let X be a topological space which is Hausdorff. Let $x \in X$. Then:*

(1) *The singleton $\{x\}$ (and hence every finite set) is always closed.*
(2) *The intersection of all open sets containing x is $\{x\}$.*

Definition. *Let (X, T) be a topological space and let $A \subset X$. Let $x \in X$.*

(1) *The **closure** of A, denoted by $\overline{A}$, is the smallest (w.r.t. "$\subset$") closed set containing A. Equivalently, it equals the intersection of all closed sets containing A.*
(2) *The **interior** of A, denoted by $\overset{\circ}{A}$, is the largest (w.r.t. "$\subset$") open set contained in A. Equivalently, it equals the union of all open sets contained in A.*
(3) *We say that x is a **limit point** of A if:*

$$\forall U \in \mathcal{V}(x) : \ U \cap A - \{x\} \neq \varnothing.$$

*The set of limit points of A is denoted by A' (we call it the **derived** set of A).*
(4) *If a point x is not a limit point, then we call it an **isolated point**.*

Remark. In some textbooks, they use the term **cluster** or an **accumulation** point instead of a limit point.

Remark. In a metric space, and in the definition of a limit point, it suffices to consider open balls of center x rather than arbitrary neighborhoods.

REMARK. Let (X, d) be a metric space and let $A \subset X$. We may easily characterize $\overset{\circ}{A}$ as follows:

$$x \in \overset{\circ}{A} \iff \exists r > 0, B(x, r) \in A.$$

We have the following relationship between the interior and the closure of a set.

PROPOSITION (for a proof, see Exercise 3.3.5). *Let X be a topological space and let $A \subset X$. Then:*

$$\overline{A^c} = (\overset{\circ}{A})^c \ and \ \overset{\circ}{A^c} = (\overline{A})^c.$$

EXAMPLES.

(1) *In usual $\mathbb{R}$, $\mathbb{Q}' = \mathbb{R}$ and $\overset{\circ}{\mathbb{Q}} = \varnothing$.*

(2) *In usual $\mathbb{R}$, if $A = [0, 2]$, then $\overset{\circ}{A} = (0, 2)$.*

(3) *In usual $\mathbb{R}$, if $A = (0, 3) \cup \{4\}$, then $\overset{\circ}{A} = (0, 3)$.*

(4) *In usual $\mathbb{R}$, $\overline{[0, 2)} = [0, 2]$.*

PROPOSITION. *Let (X, T) be a topological space and let $A \subset X$; $A_i \in X$, where I is arbitrary. Then*

(1) *A is closed iff $A = \overline{A}$.*

(2) *A is open iff $A = \overset{\circ}{A}$.*

(3)

$$x \in \overline{A} \iff \forall U \in \mathcal{V}(x): \ U \cap A \neq \varnothing.$$

(4)

$$\overline{A} = A' \cup A.$$

THEOREM. *Let (X, T) be a topological space and let $A, B \subset X$; $A_i \in X$, $i \in I$, where I is arbitrary. Then*

(1) $A \subset B \Rightarrow \overline{A} \subset \overline{B}$.

(2) $\overline{\overline{A}} = \overline{A}$.

(3) $\overline{A \cup B} = \overline{A} \cup \overline{B}$.

(4)

$$\bigcup_{i \in I} \overline{A_i} \subset \overline{\bigcup_{i \in I} A_i}.$$

(5)

$$\bigcap_{i \in I} \overline{A_i} \supset \overline{\bigcap_{i \in I} A_i}.$$

DEFINITION. *Let (X, T) be a topological space, and let $A \subset X$. We say that A is **dense** in X if $\overline{A} = X$.*

REMARK. We may also say "**everywhere dense**" in lieu of "dense".

EXAMPLES.

(1) *In usual* $\mathbb{R}$, $\mathbb{Q}$ *is dense in* $\mathbb{R}$.
(2) *In a discrete topological space* X, *the only dense set is* X *itself.*
(3) *In an indiscrete topological space, all subsets (apart from* $\varnothing$) *are dense.*

DEFINITION. *Let* X *be a topological space and let* $A \subset X$. *We say that* A *is* **nowhere dense** *in* X *if* $\overset{\circ}{\overline{A}} = \varnothing$.

EXAMPLES.

(1) *In standard* $\mathbb{R}$, $\mathbb{Z}$ *is nowhere dense in* $\mathbb{R}$ *for* $\overset{\circ}{\overline{Z}} = \varnothing$.
(2) *In standard* $\mathbb{R}$ *again, the following set*

$$\left\{ \frac{1}{n} : \ n \in \mathbb{N} \right\}$$

is also nowhere dense in $\mathbb{R}$.
(3) $\mathbb{Q}$ *is not nowhere dense in standard* $\mathbb{R}$.

DEFINITION. *Let* X *be a topological space, and let* $A \subset X$. *The* **frontier** *(also known as the* **boundary***) of* A *is (the set!) defined by:*

$$\mathrm{Fr}(A) = \overline{A} - \overset{\circ}{A}.$$

EXAMPLES.

(1) *In usual* $\mathbb{R}$,

$$\mathrm{Fr}[0,2) = [0,2] - (0,2) = \{0,2\}.$$

(2) *In a discrete topological space all frontiers are empty!*

DEFINITION. *We say that a topological space* (X, T) *is* **separable** *if it contains a* **countable** *and* **everywhere dense** *subset.*

EXAMPLE. *The standard* $\mathbb{R}$ *is separable for it contains* $\mathbb{Q}$ *which is countable and dense in* $\mathbb{R}$.

Now we give the definition of a basis for a topology.

DEFINITION. *Let* X *be a topological space. A* **basis** *(or a* **base***) for* X *is a collection* $\mathcal{B}$ *constituted of subsets of* X *verifying:*

(1) X *is a union of elements of* $\mathcal{B}$.
(2) *If* $B_1, B_2 \in \mathcal{B}$, *then* $B_1 \cap B_2$ *is a union of elements of* $\mathcal{B}$.

EXAMPLES.

(1) Let $X = \{1, 2, 3\}$ and let T be any topology on X. Set $\mathcal{B} = \{\{1, 3\}, \{2, 3\}\}$. Then $\mathcal{B}$ is not a base for T.
(2) In a metric space, the set of open balls is basis for it.
(3) Let X be a discrete topological space. Then $\mathcal{B} = \{\{x\}\}_{x \in X}$ is a basis for X.

REMARK. A given topological space may have different bases.

REMARK. There is a topology associated with a totally ordered set. It may be seen as a generalization of the usual topology of $\mathbb{R}$. We shall not consider this topology in the present book (for more details, see [10]).

Having this in mind, we set $\overline{\mathbb{R}} = \mathbb{R} \cup \{-\infty, +\infty\} = [-\infty, +\infty]$. Then we extend the order of $\mathbb{R}$ to $\overline{\mathbb{R}}$ (by staying careful with some arithmetic operations, e.g. $+\infty - \infty$ or $0 \times (\pm\infty)$, although the latter is acceptable in the context of Measure Theory). The set $\overline{\mathbb{R}}$ equipped with this order and the associated topology is called **the extended real line**.

3.1.2. The Subspace Topology.

DEFINITION. Let (X, T) be a topological space, and let $A \subset X$. Set

$$T_A = \{A \cap U : U \in T\}.$$

Then (A, T_A) is a topology on A (for a proof see Exercise 3.3.16). We call it the **subspace topology**.

It may also be called **relative** or **induced** topology.

REMARK. Going back to the definition of an isolated point, combined with the subspace topology we may state with ease that: x is an isolated point of a A ($A \subset X$, X is a topological space) iff $\{x\}$ is open in A as a subspace of X.

REMARK. The set $A = [0, 1)$ is not open in usual $\mathbb{R}$, but it is open in T_A. So one has to be careful when using the word "open" when dealing with the subspace topology. However, there are cases when the "two" open sets coincide. We have

PROPOSITION. Let X be a topological space, and let $A \subset X$ be equipped with the subspace topology. Then if U is open in A and A is open in X, then U is open in X.

The next result tells us what closures and closed sets are in the subspace topology.

PROPOSITION. *Let X be a topological space and let $A \subset X$ be endowed with the subspace topology. Then:*

   (1) *The closed sets in A are of the form $V \cap A$ where V is a closed set in X.*

   (2) *The closure of B in A is $\overline{B} \cap A$ where $\overline{B}$ is the closure of B in X.*

3.1.3. The Product and Quotient Topologies.

DEFINITION. *Let X and Y be two topological spaces. We call an **elementary open** of $X \times Y$ every set of the form $U \times V$, where U is open in X and V is open in Y.*

DEFINITION. *Let X and Y be two topological spaces. Let $\mathcal{B}$ be the collection of elementary open sets in $X \times Y$.*

*The **product topology** on $X \times Y$ is the topology having $\mathcal{B}$ as a basis.*

REMARK. An open set in a product space is not necessarily an elementary open! For a counterexample, see the "True or False" section.

THEOREM (for a proof, see Exercise 3.3.34). *Let X and Y be two topological spaces. Let A and B two subsets of X and Y respectively. Then*

   (1) $\overline{A \times B} = \overline{A} \times \overline{B}$.

   (2) $\overset{\circ}{\overbrace{A \times B}} = \overset{\circ}{A} \times \overset{\circ}{B}$.

We finish this section with the quotient topology.

PROPOSITION (for a proof, see Exercise 3.3.37). *Let X be a topological space and let R be an equivalence relation on X. Let $\varphi : X \to X/R$ be the quotient map. Let*

$$T = \{A \in X/R : \varphi^{-1}(A) \text{ is open in } X\}.$$

Then T is a topology in X/R.

DEFINITION. *The topology defined on X/R in the previous proposition is called the **quotient topology**.*

3.2. True or False: Questions

QUESTIONS. Comment on the following questions/statements and indicate those which are false and those which are true when this applies. Justify your answers.

   (1) Let X be a non-empty set. We can always define a topology on X.

(2) To prove a given part, T say, is a topological space we need to show (*among others*) that the finite intersection of elements in T is again in T. But it is always sufficient to do that for the intersection of two elements. Why is this?

(3) Let T and T' two topologies on the same set X. Then $T \cup T'$ is a topology on X.

(4) Let T and T' two topologies on the same set X. Then $T \cap T'$ is a topology on X.

(5) Let X be a topological space and let $A \subset X$. The mappings $A \mapsto \overline{A}$ and $A \mapsto \overset{\circ}{A}$ are "monotonic" with respect to "$\subset$".

(6) Let X be a topological space and let $A \subset X$. Then it may well happen that $\overset{\circ}{A} = X$.

(7) Let X be a topological space. Let $A \subset X$. If A is dense in X, then $\overset{\circ}{\overline{A}} = \varnothing$.

(8) Let X be a topological space and let $A \subset X$. Then it may well happen that $\overline{A} = \varnothing$.

(9) Let X be a topological space. Whether we say X is separable or separated, we are talking about the same thing!

(10) Let X be a topological space and let A be a topological subspace of X. Then

$$A \text{ Hausdorff } \Longrightarrow \overline{A} \text{ Hausdorff.}$$

(11) Every subspace of a Hausdorff space is Hausdorff.

(12) Let T and T' be two topologies on X such that T' is finer than T. If T is Hausdorff, then so is T'.

(13) Every subspace of a separable space is separable.

(14) There exists a topological space X in which all subsets (except $\varnothing$) are dense.

(15) Let X be a topological space and let Y be a subset of X endowed with the induced topology of X. Then

$$\overset{\circ Y}{A} = Y \cap \overset{\circ X}{A}.$$

(16) There is a definition of a bounded set in an arbitrary topological space.

(17) Let X and Y be two given sets and let A and B be two subsets of X and Y respectively. Then

$$(A \times B)^c = A^c \times B^c.$$

(18) Let X and Y be two topological spaces. If $\Omega = U \times V$ is open (respectively closed) in $X \times Y$, then U is open (respectively closed) in X and V is open (respectively closed) in Y.

(19) Let X and Y be two topological spaces. We endow $X \times Y$ with the product topology. Then there is an open set in $X \times Y$ which is not of the form $U \times V$ where U is open in X and V is open in Y.

(20) Let X be a topological space and let $A \subset X$. If we unify the interior of A together with its exterior we get back the whole of X.

(21) Let X be a topological space and let $A \subset X$. Then $\overset{\circ}{A} = \overline{\overset{\circ}{A}}$.

(22) Let X be a topological space and let $A \subset X$. Then $\overline{A} = \overset{\circ}{\overline{A}}$.

(23) Let (X, d) be a metric space. Let A be a closed set in X. Every point in A is a limit point of A.

(24) Let X be a topological space and let $A \subset X$. Denote the derived set of A, i.e. the set of limit points of A, by A'. Then A' is always closed.

(25) Let X be a topological space and let $A \subset X$. Then A and A' can be disjoint, comparable and they can be equal as well.

(26) Let X be a topological space and let A and B be two subsets of X. Then
$$A \subset B \iff A' \subset B'.$$

(27) Let X be a topological space and let $A \subset X$. Then
$$\mathrm{Fr}(\overline{A}) = \mathrm{Fr}(A).$$

(28) Let X be a topological space and let $A \subset X$. Then
$$\mathrm{Fr}(\overset{\circ}{A}) = \mathrm{Fr}(A).$$

(29) Let X be a topological space and let $A, B \subset X$. Then
$$\mathrm{Fr}(A \cup B) = \mathrm{Fr}(A) \cup \mathrm{Fr}(B).$$

(30) Let X be a topological space and let $A \subset X$. Then
$$\mathrm{Fr}\, A \subset A.$$

(31) Let X be a topological space and let $A \subset X$. Then
$$\mathrm{Fr}\, A \subset \overline{A}.$$

(32) Let X be a topological space and let $A, B \subset X$ such that $A \subset B$. Then
$$\mathrm{Fr}(A) \subset \mathrm{Fr}(B).$$

(33) Let (X, d) be a metric space and let $x \in X$. Let: $B(x, r)$ be the open ball, $B_c(x, r)$ be the closed ball and $S(x, r)$ be the sphere, all of radius $r > 0$ and center x. Then
$$\mathrm{Fr}(B(x, r)) = S(x, r) \text{ or } \mathrm{Fr}(B_c(x, r)) = S(x, r).$$

(34) In usual $\mathbb{R}$, the sets

$$(-1,1], \ (-2,2), \ (0,2], \ [-1,0] \text{ and } [0,1]$$

are neighborhoods of 0.

(35) Let T and T' be two topologies on the same set X. If $T \subset T'$, then every open set in T is so in T'. For closed sets, every closed set in T' is so in T.

(36) The quotient of a separated space is separated.

3.3. Exercises With Solutions

Exercise 3.3.1. List all possible topologies which can be defined on the sets $X = \{1\}$ and $Y = \{1,2\}$.

Exercise 3.3.2. Let $X = \{a,b,c,d,e\}$. We define the subset T of $\mathcal{P}(X)$ as $T = \{\varnothing, \{a\}, \{c,d\}, \{a,c,d\}, \{b,c,d,e\}, X\}$.

(1) Show that T is a topology on X.
(2) What are the closed sets in T?
(3) What are the closures and the interiors of $\{a\}$, $\{b\}$ as well as their boundaries?
(4) Give the closure of $\{a,b\}$. What can deduce from that?
(5) Give the neighborhoods of c and d.
(6) Is T Hausdorff?

Exercise 3.3.3. Let $n \in \mathbb{N}$. Set $A_n = \{n, n+1, n+2, \cdots\}$ and

$$T = \{\varnothing, A_n\}_{n \in \mathbb{N}}.$$

(1) Verify that T is a topology on $\mathbb{N}$.
(2) Is $\{1,3,5,7,\cdots\}$ open in T?
(3) Determine $\mathcal{V}(2)$ and $\mathcal{V}(3)$.
(4) Is T separated?
(5) What are the closed sets in T?
(6) What is the interior and the closure of $\{4\}$ and of $\{2,4,6,8,\cdots\}$?
(7) What are the dense sets in T?

Exercise 3.3.4. Let T be a topology on the set $X = \{a,b,c\}$. Show that if the singletons $\{a\}$, $\{b\}$ and $\{c\}$ are open in T, then T is the discrete topology.

Exercise 3.3.5. Let (X,T) be a topological space and let $A \subset X$. Show that

$$\overline{A^c} = (\mathring{A})^c \text{ and } \left(\overline{A}\right)^c = (\mathring{A^c}).$$

Exercise 3.3.6. Let X be a topological space. Show that for every subset A of X and for every open set U we have

$$A \cap U = \varnothing \Leftrightarrow \overline{A} \cap U = \varnothing.$$

Exercise 3.3.7. Let $\mathbb{R}$ be endowed with its standard topology.
(1) Find the closures of $\mathbb{Q}$, $\mathbb{R} \setminus \mathbb{Q}$, $(0,1]$ and $\{1\} \cup (2,3]$.
(2) Find the interior of $\mathbb{Q}$, $\mathbb{R} \setminus \mathbb{Q}$ and $(0,1]$.
(3) Do we always have

$$\left(\bigcap_{n \geq 1} A_n \right)^{\circ} = \bigcap_{n \geq 1} \overset{\circ}{A}_n \text{ or } \overline{\bigcup_{n \geq 1} A_n} = \bigcup_{n \geq 1} \overline{A_n}?$$

(4) Is $\mathbb{R}$ Hausdorff?

Exercise 3.3.8. Let (X, d) be a metric space and let $A \subset X$ be an open set. Show that every point in A is a limit point of A.

Exercise 3.3.9. Let $\mathbb{R}$ be endowed with its usual topology and let $A = \{\frac{1}{n} : n \geq 1\}$.
(1) Find the interior, the isolated and the limit point(s) of A.
(2) Find $\overline{A}$ and check that A is not closed. What is then the frontier of A.
(3) Show that A is nowhere dense in $\mathbb{R}$.

Exercise 3.3.10. Give an example of a set A for which the sets A, $\overset{\circ}{A}$, $\overline{A}$, $\overline{\overset{\circ}{A}}$ and $\overset{\circ}{\overline{A}}$ are pairwise different.

Exercise 3.3.11. Let X be a non-void set. Equip it with the discrete topology and let $A \subset X$. What is A' worth?

Exercise 3.3.12. Let $\mathbb{R}$ be equipped with its usual topology. Let A be a non-empty and bounded subset of $\mathbb{R}$.
(1) Show that $\inf A$, $\sup A \in \overline{A}$.
(2) Does this result remain valid in other topologies on $\mathbb{R}$?
(3) Do we have an analogous result for $\overset{\circ}{A}$?

Exercise 3.3.13. Let A be a non-void and bounded above subset of $\mathbb{R}$.
(1) Show that in usual $\mathbb{R}$, $\sup A = \sup \overline{A}$.
(2) Show that the previous result need not hold in another topology of $\mathbb{R}$.
(3) In the usual topology of $\mathbb{R}$ again, do we have $\sup \overset{\circ}{A} = \sup A$?

Exercise 3.3.14. Let (X, d) be a metric space. Let $r > 0$ and let $x \in X$. We denote the open ball in X of center x and radius r by $B(x, r)$ while the closed ball in X of center x and radius r is denoted by $B_c(x, r)$.

(1) Show that $\overline{B(x, r)} \subset B_c(x, r)$.

(2) Give two examples (one in $\mathbb{R}$ and one in an arbitrary metric space) which show that the backward inclusion is not always true.

Exercise 3.3.15. Let X be a non-empty set and let (Y, T) be a topological space. Let $f : X \to Y$ be some function. Show that

$$T' = \{f^{-1}(U) : U \in T\}$$

is a topology on X.

Exercise 3.3.16. Consider a non-empty set X with a topology T. Let $A \subset X$. Show that

$$T_A = \{A \cap U : U \in T\}$$

is a topology on A.

Exercise 3.3.17. Let $\mathbb{R}$ be endowed with its standard topology. Let A be a topological subspace of $\mathbb{R}$.

(1) Is $\{3\}$ open in $A = [0, 1) \cup \{3\}$?

(2) Are $[0, 1)$ and $(0, 1)$ open in $A = [0, 1]$?

(3) Let $n \in \mathbb{N}$. Is $\{n\}$ open in $A = \mathbb{N}$?

(4) Show that $[0, 1]$ and $(2, 3)$ are both open in $A = [0, 1] \cup (2, 3)$. What can you deduce from that?

(5) What is the closure of $\left(0, \frac{1}{2}\right)$ in $A = (0, 1]$?

(6) Go back to Exercise 3.3.2 and set $A = \{b, c, d\}$. Give the subspace topology on A. Give the closure of $\{b, d\}$ in A by two methods (you will need to find its closure in X as well).

Exercise 3.3.18. Endow both $A = \mathbb{Q}$ and $B = \mathbb{R} \setminus \mathbb{Q}$ with the induced usual topology of $\mathbb{R}$.

(1) Is $X = A \cap [\sqrt{2}, \pi]$ clopen in A?

(2) Is $Y = B \cap [0, 2]$ clopen in B?

(3) Is $Z = A \cap [\sqrt{2}, \pi)$ closed or open in A?

(4) Is $Z' = B \cap [\sqrt{2}, \pi)$ closed or open in B?

Exercise 3.3.19. Show that in $\mathbb{R}$, $[0, 1] \cap \mathbb{Q}$ is dense in $[0, 1]$.

Exercise 3.3.20. Consider the set

$$X = \left\{ x \in [0, 1] : x = \sum_{n=1}^{\infty} \frac{\alpha_n}{10^n} \text{ where } \alpha_n = 3 \text{ or } \alpha_n = 5 \right\}.$$

(1) Describe the elements of X.
(2) Show that X is not dense in $[0, 1]$.

Exercise 3.3.21. Let X a set with card$X \geq 2$. Endow X with the indiscrete topology. Show that X is not metrizable using different approaches.

Exercise 3.3.22 (Co-finite Topology). Let X be an infinite set and let T be the family given by

$$T = \{\varnothing\} \cup \{U \subset X : U^c \text{ finite }\}.$$

This will be called in the sequel the **Co-finite topology**. It is a particular case of more general topology called the **Zariski topology**. In this exercise X is taken to be $\mathbb{R}$ except for the last equation.

(1) Show that T is a topological space in $\mathbb{R}$.
(2) Describe the closed sets in this topology.
(3) Compare T with the usual topology.
(4) Is it Hausdorff?
(5) Is it metrizable?
(6) (a) Let A be a finite subset of $\mathbb{R}$. Find $\overline{A}$, $\overset{\circ}{A}$ and the boundary of A.
    (b) The same questions if we assume that A is infinite.
(7) Is $\mathbb{R}$ separable with respect to T?
(8) What does T become if X is a finite set?

Exercise 3.3.23. Is $T = \{\varnothing, \mathbb{R}\} \cup \{U \subset \mathbb{R} : U^c \text{ infinite}\}$ a topology on $\mathbb{R}$?

Exercise 3.3.24 (co-countable topology). Let X be an uncountable set. Put

$$T = \{\varnothing\} \cup \{U \subset X : U^c \text{ is countable}\}.$$

(1) Show that T a topological space on X (called the **co-countable topology**).
(2) Is it Hausdorff?
(3) Let A be a *proper closed* set of X. Show that all subsets of A are closed.
(4) Prove that the countable intersection of open sets in T remains open. Is this result true in usual $\mathbb{R}$?
(5) Show that the finite intersection of open sets is non-empty. Is this result true in usual $\mathbb{R}$?
(6) Is $\mathbb{Q}$ dense in $\mathbb{R}$ endowed with the co-countable topology? What about $\mathbb{R} \setminus \mathbb{Q}$ or $[0, 1]$?
(7) Give an example of a set X so that T reduces to the discrete topology.

Exercise 3.3.25. Let X be a non-empty set. Let $a \in X$ be fixed and set

$$T = \{\varnothing\} \cup \{U \subset X : a \in U\}.$$

(1) Check that T is a topology on X.
(2) Is T Hausdorff?
(3) Find $\{a\}'$.
(4) Deduce that open sets (except $\varnothing$) are all dense in T.
(5) Let A be a set containing a. Prove that $A' = X - \{a\}$.
(6) Is X separable? What about $X - \{a\}$?
(7) Show that every proper subset of X is nowhere dense.
(8) What is $T_{\{a\}^c}$, the induced topology on $\{a\}^c$? Is it Hausdorff?

Exercise 3.3.26. Let $a > 0$. Let $X = [-a, a]$. We declare

$$U \text{ "open" in } T \iff \{0\} \not\subset U \text{ or } (-a, a) \subset U.$$

(1) Show that T is actually a topology on X.
(2) Give all the closed sets in T.
(3) Set $A = \{\frac{a}{3}\}$. What is $\overline{A}$?
(4) Show that 0 is a limit point of any subset of $B \subset (-a, a)$.

Exercise 3.3.27. Let $X = [0, 2)$. Define

$$T = \{[0, a) : 0 \le a \le 2\}.$$

(1) Show that T is a topology on X.
(2) Give an example which shows that the arbitrary intersection of elements of T need not be in T again.
(3) Is T Hausdorff?
(4) What are the closed sets in T?
(5) Show that the closure of $A = [1, \frac{3}{2}]$ in T is $[1, 2)$ and that its interior is the empty set.
(6) Is X separable with respect to T?

Exercise 3.3.28. Let T be a topology on X. Assume that T is Hausdorff and let $x \in X$.

(1) Show that $\{x\}$ (and hence every finite set) is always closed.
(2) Show, by an example, that the separation hypothesis cannot be dispensed with.
(3) Give an example of a *non-separated* space in which singletons are closed.
(4) Show that the intersection of all open sets containing x is $\{x\}$.
(5) Give an example of a *non-separated* space in which the previous result holds.

Exercise 3.3.29 (Lower Limit Topology). Set

$$\mathcal{B} = \{[a, b) : \ a, b \in \mathbb{R}\}.$$

The topology generated by $\mathcal{B}$ is called the **lower limit topology** and it is denoted by $\mathbb{R}_\ell$ (it is called in some references the **Sorgenfrey Topology**).

(1) Check that $\mathcal{B}$ is in effect a base for $\mathbb{R}$.
(2) Show that $\mathbb{R}$ is strictly coarser than $\mathbb{R}_\ell$.
(3) Give examples of closed, open and clopen sets. What is the nature of $(a, b]$?
(4) Is $\mathbb{R}_\ell$ Hausdorff?
(5) Is $\mathbb{R}_\ell$ separable?

REMARK. Likewise, we may define a topology on $\mathbb{R}$ using the base $\mathcal{B} = \{(a, b] : \ a, b \in \mathbb{R}\}$. This new topology is called the **upper limit topology**.

Exercise 3.3.30 (K-topology). Set $K = \{\frac{1}{n} : \ n \in \mathbb{N}\}$. The K-*topology* on $\mathbb{R}$ is generated by the basis

$$\mathcal{B} = \{(a, b)\} \cup \{(a, b) - K\}$$

where $a, b \in \mathbb{R}$. It is denoted by $\mathbb{R}_K$.

(1) Show that $\mathbb{R}_K$ is strictly finer than $\mathbb{R}$.
(2) Is $\mathbb{R}_K$ Hausdorff?
(3) Are $\mathbb{R}_\ell$ and $\mathbb{R}_K$ comparable?
(4) Is K closed in $\mathbb{R}_K$?
(5) Calculate K'.

Exercise 3.3.31.

(1) Show that $\mathcal{B} = \{(a, b) : \ a, b \in \mathbb{Q}\}$ is a basis generating the usual $\mathbb{R}$.
(2) Show that $\mathcal{B}' = \{[a, b) : \ a, b \in \mathbb{Q}\}$ is a basis that generates a topology, denoted by $\mathbb{R}_{\mathcal{B}'}$, different from $\mathbb{R}_\ell$.

Exercise 3.3.32. Find

$$d((0, 1) \cap \mathbb{Q}) \text{ and } d((0, 1) \cap \mathbb{R} \setminus \mathbb{Q}).$$

Exercise 3.3.33. Let (X, d) be a metric space and let A be a non-empty subset of X. Let $x \in X$. Define

$$d(x, A) = \inf_{a \in A} d(x, a).$$

(1) Justify the existence of $d(x, A)$.

(2) Show that

$$d(x, A) = 0 \Leftrightarrow x \in \overline{A}.$$

Deduce yet another proof that singletons are closed in metric spaces (cf. Exercise 3.3.28).
(3) Show that $d(x, A) = d(x, \overline{A})$.

Exercise 3.3.34. Let X and Y be two topological spaces. Let A and B two subsets of X and Y respectively.

(1) Show that $\overline{A \times B} = \overline{A} \times \overline{B}$.
(2) Show that

$$\overset{\circ}{\overbrace{A \times B}} = \overset{\circ}{A} \times \overset{\circ}{B}.$$

Exercise 3.3.35. Show that the finite product of separable spaces is separable.

Exercise 3.3.36. On usual $\mathbb{R}^2$, consider the sets

$$A = \left\{ (x, y) : \ y = \sin\frac{1}{x}, x > 0 \right\}, \ B = \{(x, x) : \ x \in \mathbb{R}\},$$

$$C = \{(x, y) \in \mathbb{R}^2 : \ |x| < 2, \ |y| < 3\} \text{ and } D = \{(1, 1)\} \times C$$

(D being defined on usual $\mathbb{R}^4$). Determine the interior and the closure of each of these four sets.

Exercise 3.3.37. Let X be a topological space and let R be an equivalence relation on X. Let $\varphi : X \to X/R$ be the quotient map. Set

$$T = \{A \in X/R : \ \varphi^{-1}(A) \text{ is open in } X\}.$$

Show that T is a topology in X/R.

Exercise 3.3.38. Equip $\mathbb{R}$ with its usual topology. Define a relation $\mathcal{R}$ on $\mathbb{R}$ by

$$x\mathcal{R}y \iff x - y \in \mathbb{Q}$$

for all $x, y \in \mathbb{R}$.

(1) Check that $\mathcal{R}$ is indeed an equivalence relation on $\mathbb{R}$.
(2) Show that the quotient topology $\mathbb{R}/\mathbb{Q}$ is not Hausdorff (cf. Exercise 4.5.15).
(3) Show that $\mathbb{R}/\mathbb{Q}$ is in fact the discrete topology.

3.4. Tests

Test 7. Let (X, T) be a topological space. Let $A \subset X$. Assume that
$$\forall x \in A, \exists U \in T : \ x \in U \subset A.$$
(1) Show that A is open in (X, T).
(2) Does this result remind you of something you are familiar with?

Test 8. Is the set $A = \{1, 2^2, 3^2, \cdots\}$ closed in $\mathbb{R}$?

Test 9. Does every closed set in $\mathbb{R}$ intersect $\mathbb{Q}$?

Test 10. Is the function assigning to each set its closure one-to-one?

Test 11. Let $X = \mathbb{N} - \{1\}$. Define $A_n = \{d \in X : \ d \mid n\}$ for $n \in \mathbb{N}$ where $d \mid n$ stands for d divides n. Is $T = \{A_n\}_{n \in \mathbb{N}}$ a topology on X?

Test 12. Let $X = \{a, b, c\}$. Find the topologies on X having exactly four open sets.

Test 13. Give an example of a bounded set having five limit points.

Test 14. Let X be a topological space. Let $A \subset X$. Find A' in the case of the the discrete topology and in the case of the co-finite topology.

Test 15. Is the topology of Exercise 3.3.27 metrizable?

Test 16. Are the co-countable (see Exercise 3.3.24) and the usual topologies comparable on $\mathbb{R}$?

Test 17. Is $\mathbb{R}$ separable when endowed with the discrete topology?

Test 18. Let
$$T = \{(-a, a) : \ a \geq 0\} \cup \{\mathbb{R}\}.$$
(1) Check that T is a topology on $\mathbb{R}$.
(2) Find the interior and the closure of $[-1, 2]$.
(3) The same question for $\{0\}$ and $\{1\}$.

Test 19. Let $A \subset X$ where X has the topology of Exercise 3.3.25. When does A inherits the discrete topology?

Test 20. Let X be a topological space and let $A \subset X$. Can $\mathrm{Fr}(A)$ be equal to some open set in X?

Test 21. Is $\{1, \frac{1}{2}, \frac{1}{3}, \cdots\} \cup \{0\}$ closed in $\mathbb{R}_K$?

Test 22. Show that $\{(x, y) \in \mathbb{R}^2 : \ xy < 2\}$ is open in usual $\mathbb{R}^2$ but it is not an open ball.

3.5. More Exercises

Exercise 3.5.1. Let X be a non-empty set and let A, B be two non-void proper subsets of X. Let $T = \{\varnothing, A, B, X\}$.

(1) Is T always a topology on X?

(2) What are the conditions so that T becomes a topology on X?

Exercise 3.5.2. Do all the questions of Exercise 3.3.2 for the couple (X, T) where $X = \{a, b, c, d\}$ and

$$T = \{\varnothing, \{a\}, \{b\}, \{a, b\}, \{a, b, d\}, \{a, b, c\}, X\}.$$

Exercise 3.5.3. Let $n \in \mathbb{Z}^+$.

$$T = \{\varnothing\} \cup \{n\mathbb{Z}\}.$$

Show that T is not a topology on $\mathbb{Z}$.

Exercise 3.5.4. Let X be an indiscrete topological space and let $A \subset X$. Find A'.

Exercise 3.5.5. Let A be some subset of a topological space X. Show that $A' = \overline{A}'$ where the "'" is for the derived set.

Exercise 3.5.6. Find $\overset{\circ}{\mathbb{Q}}$ in $\mathbb{R}$ the following cases:

(1) $\mathbb{R}$ equipped with the co-finite topology;

(2) $\mathbb{R}$ endowed with the co-countable topology.

Exercise 3.5.7. Let $T \subset \mathcal{P}(\mathbb{N})$ be such that $U \subset T$ is "open" iff $U = \varnothing$ or $1 \in U$.

(1) Verify that T is indeed a topology on $\mathbb{N}$.

(2) Check that $\{1\}$ is dense in $\mathbb{N}$.

Exercise 3.5.8. What is the topology generated by the intervals of the form $[x, x + 1]$ where $x \in \mathbb{R}$?

Exercise 3.5.9. Let A be a non-void and bounded below subset of $\mathbb{R}$.

(1) Show that in usual $\mathbb{R}$, $\inf A = \inf \overline{A}$.

(2) Show that the previous result need not hold in another topology of $\mathbb{R}$.

(3) In the usual topology of $\mathbb{R}$ again, do we have $\inf \overset{\circ}{A} = \inf A$?

(4) Give an example of a topological space in which $\inf \overset{\circ}{A} = \inf A$.

Exercise 3.5.10. Show that a metric space is Hausdorff.

Exercise 3.5.11. Let X be a set endowed with the discrete topology. Let $A \subset X$.

(1) Find $\overline{A}$, $\overset{\circ}{A}$ and the boundary of A.
(2) What are the dense parts of X?
(3) When is X separable?

Exercise 3.5.12. Set

$$T = \{\varnothing, \mathbb{R}\} \cup \{(a, \infty)\}_{a \in \mathbb{R}}.$$

(1) Show that T is a topology on $\mathbb{R}$.
(2) Is T Hausdorff?
(3) List the closed sets in T.
(4) What are the closures of the sets $\{0\}$, $\{0, 5, 11\}$ and $[2, 8)$.
(5) What is the interior of the interval $[0, \infty)$.
(6) Would T have remained a topology on $\mathbb{R}$ have we taken $a \in \mathbb{Q}$?

Exercise 3.5.13. Let $A = [-2, 2]$ considered as a topological subspace of $\mathbb{R}$. Which of the following sets are open in A or in $\mathbb{R}$?

(1) $B = \{x \in \mathbb{R} : 1 < |x| < 2\}$;
(2) $C = \{x \in \mathbb{R} : 1 < |x| \leq 2\}$;
(3) $D = \{x \in \mathbb{R} : 1 \leq |x| < 2\}$?

Exercise 3.5.14. We say that a topological space T on some set X is T_1 if

$$\forall x, y \in X; x \neq y, \exists U \in T \text{ with } x \in U \text{ and } y \notin U.$$

(1) Show that all Hausdorff spaces are T_1.
(2) Show that the co-finite topology on $\mathbb{R}$ is T_1 (but not Hausdorff).
(3) Is $\mathbb{R}$ T_1 with respect to:
    (a) the countable complement topology?
    (b) the topology of Exercise 3.3.25?
    (c) the topology of Exercise 3.3.26?

Exercise 3.5.15. We say that a topological space T on some set X is **first countable** if it has a countable neighborhood basis at each of its points. We say that T is **second countable** if it has a countable basis.

(1) Show that any second countable spaces is necessarily a first countable one. What about the converse? (hint: consider $\mathbb{R}$ equipped with the lower limit topology).
(2) Show that a second countable space is separable. Is the converse true? (hint: consider again $\mathbb{R}$ equipped with the lower limit topology).
(3) Show that the usual topology and more generally, metric spaces, are first countable.

(4) Show that $\mathbb{R}$ is not first countable in the co-finite topology.
(5) Are the topologies of: Exercise 3.3.25, Exercise 3.3.26 first countable? second countable?

Exercise 3.5.16. We say that a number is **dyadic** if it is a rational written as $\frac{m}{2^n}$, $m \in \mathbb{Z}$ and $n \in \mathbb{N}$. Show that the set of dyadic numbers is dense in $\mathbb{R}$.

Exercise 3.5.17. Let X be a topological space. Let A be an open subset of X and let $B \subset X$. Show that
(1) If $A \cap B = \varnothing$, then $A \cap \overline{B} = \varnothing$.
(2) If B is dense in X, then $\overline{A \cap B} = \overline{A}$.
(3) Deduce that if A and B are both dense in X, then so are $A \cup B$ and $A \cap B$.

Exercise 3.5.18. Give $\overline{A}$ and $\overset{\circ}{A}$ where

$$A = \left\{ \frac{1}{n} : n \in \mathbb{N} \right\}$$

with respect to each of the following topologies:
(1) the co-finite topology,
(2) the co-countable topology,
(3) the lower limit topology,
(4) the topology of Exercise 3.3.26 with $a = 1$,
(5) the topology of Test 18.

Exercise 3.5.19. Let X be a topological space. Let $A \subset X$.
(1) Show that

$$A \text{ is clopen } \iff \operatorname{Fr}(A) = \varnothing.$$

(2) Is $\operatorname{Fr}(A)$ always nowhere dense?
(3) Show that $\operatorname{Fr}(A)$ is nowhere dense whenever A is either open or closed.

Exercise 3.5.20. Let X be a topological space and let $A \subset X$.
(1) Show that A is nowhere dense iff $\overline{A}^c$ is dense in X.
(2) Does the arbitrary union of nowhere dense sets remain dense?
(3) Show that the *finite* union of nowhere dense sets remains dense.

Exercise 3.5.21. Let X be a topological space. If A and B are nonvoid elements of X such that $A \cap B = \varnothing$, then show that

$$\overset{\circ}{\overline{A}} \cap \overset{\circ}{\overline{B}} = \varnothing.$$

Exercise 3.5.22. Let us endow $\mathbb{R}^2$ with the metric defined, for all $x = (x_1, x_2)$ and $y = (y_1, y_2)$, by

$$d_\infty(x, y) = \max(|x_1 - y_1|, |x_2 - y_2|).$$

Let $B_c((0,0), 1)$ be the closed ball in $\mathbb{R}^2$. Let $A = \{(z, z) : z \in \mathbb{R}\}$.

(1) Show that $\overset{\circ}{\overbrace{B_c((0,0), 1)}} = B((0,0), 1)$.

(2) Show that $\overset{\circ}{A} = \varnothing$.

Exercise 3.5.23. Let $A \subset \mathbb{R}$.

(1) Show that if in usual $\mathbb{R}$ we have $\overset{\circ}{A} \neq \varnothing$, then A must be uncountable.

(2) Is this result always true in other topologies on $\mathbb{R}$?

Exercise 3.5.24. On $\mathbb{R}^2$, consider the collection $(B(0, r))_r$ where r is a positive number allowed to be $+\infty$ as well.

(1) Show that the collection $(B(0, r))_r$ defines a topology on $\mathbb{R}^2$.

(2) Is this topology Hausdorff?

(3) Is this topology finer than the standard one on $\mathbb{R}^2$?

Exercise 3.5.25. Show that the product of two separated spaces is separated.

Exercise 3.5.26.

(1) Show that the product of two discrete spaces is discrete.

(2) Show that the product of two indiscrete spaces is indiscrete.

Exercise 3.5.27.

(1) Show that usual $\mathbb{R} \setminus \mathbb{Q}$ is separable.

(2) Show that usual $\mathbb{C}$ is separable.

Exercise 3.5.28. Let (X, d) be a metric space. Let $A \subset X$ be bounded.

(1) Is $d(A) = d(\overline{A})$?

(2) Is $d(A) = d(\overset{\circ}{A})$?

CHAPTER 4

Continuity and Convergence

4.1. What You Need to Know

4.1.1. Continuity.

DEFINITION. *Let (X, T) and (Y, T') be two topological spaces. We say that the function $f : (X, T) \to (Y, T')$ is **continuous** if*

$$\forall U \in T', \ f^{-1}(U) \in T.$$

EXAMPLES.

(1) *Let (X, T) be a topological space let $f : (X, T) \to (X, T)$ be a function defined by $f(x) = x$. Then f is continuous.*
(2) *Let (X, T) and (Y, T') be two topological spaces let $f : (X, T) \to (Y, T')$ be a function defined by $f(x) = a$, where $a \in Y$. Then f is continuous.*

PROPOSITION. *Let (X, T), (Y, T') and (Z, T'') be three topological spaces. Let $f : (X, T) \to (Y, T')$ and $g : (Y, T') \to (Z, T'')$ be two continuous functions. Then $g \circ f : (X, T) \to (Z, T'')$ is continuous.*

The next results gives a characterization of continuity:

THEOREM (cf. Exercise 4.3.3). *Let (X, T) and (Y, T') be two topological spaces. Let $f : (X, T) \to (Y, T')$ be a function. Then the following are equivalent:*

(1) *f is continuous;*
(2) *for any $A \subset X$, $f(\overline{A}) \subset \overline{f(A)}$;*
(3) *for any closed element V in T', $f^{-1}(V)$ is closed in T.*

DEFINITION. *Let (X, T) and (Y, T') be two topological spaces. We say that the bijective mapping $f : (X, T) \to (Y, T')$ is a **homeomorphism** if f and f^{-1} are both continuous.*

*We say that two topological spaces are **homeomorphic** if there exists a homeomorphism between them.*

45

EXAMPLES.

(1) *Let $f : \mathbb{R} \to (-1, 1)$ be defined by:*
$$f(x) = \frac{x}{1 + |x|}.$$

Then f is a homeomorphism (see Exercise 4.3.7).

(2) *Every isometry is a homeomorphism.*

(3) *Any two open intervals in standard $\mathbb{R}$ are homeomorphic (see Exercise 4.3.7).*

REMARK. Let X and Y be two topological spaces. Let $f : X \to Y$ be a function. Then it is clear that f is a homeomorphism iff the following three conditions hold:

(1) f is a bijection;

(2) for any open U in Y, $f^{-1}(U)$ is open in X;

(3) for any open V in X, $f(V)$ is open in Y.

DEFINITION. *Let X and Y be two topological spaces. Let $f : X \to Y$ be a function. Then*

(1) *f is said to be **open** if for any open set U in X, $f(U)$ is open in Y.*

(2) *f is said to be **closed** if for any closed set V in X, $f(V)$ is closed in Y.*

PROPOSITION. *An open, continuous and bijective map is a homeomorphism.*

DEFINITION. *A **topological property** is a property shared by a given topological space and any other topological space homeomorphic to it.*

EXAMPLES.

(1) *We saw above that $\mathbb{R}$ is homeomorphic to $(-1, 1)$. Thus two consequences arise:*

(a) *Boundedness is not a topological property (see also Exercise 4.3.5);*

(b) *The length is not a topological property either*

(2) *Closedness is not a topological property either. In the usual topology, the closed $\mathbb{R}$ is homeomorphic to the open $\mathbb{R}_+^*$ via the function "ln".*

REMARK. Throughout this book, many examples of topological properties will be met.

DEFINITION. *Let $f : X \to Y$ be a function where X and Y are two topological spaces. The **graph** of f, denoted by G_f, is defined as:*

$$G_f = \{(x, f(x)) : \ x \in X\}.$$

4.1.2. Convergence.

DEFINITION. *Let (X, T) be a topological space. A sequence (x_n) in X **converges** to $x \in X$ if:*

$$\forall U \in \mathcal{V}(x), \ \exists N \in \mathbb{N}, \forall N \in \mathbb{N} \ (n \geq N \implies x_n \in U).$$

EXAMPLE. *Let $X = \{1, 2, 3\}$ be equipped with the topology:*

$$T = \{\varnothing, \{2\}, \{1, 2\}, \{2, 3\}, X\}.$$

Then the sequence defined by $x_n = 2$ converges to 2. It also converges to 1 and 3. Thus a sequence in an arbitrary topological space can converge to more than one limit.

REMARK. One has to be careful with sequences in topological spaces. The topology which equips a set may give us "surprises". For example, sequences like $x_n = n$ or $(-1)^n$ may converge.

In a metric space we have the following definition (equivalent to the previous one in this setting)

DEFINITION. *Let (X, d) be a metric space. A sequence (x_n) in X converges to $x \in X$ if:*

$$\forall \varepsilon > 0, \ \exists N \in \mathbb{N}, \forall N \in \mathbb{N} \ (n \geq N \implies d(x_n, x) < \varepsilon.).$$

EXAMPLE. *In a discrete metric space, the only convergent sequences are the eventually constant ones.*

THEOREM. *In a separated (Hausdorff) topological space, a sequence cannot converge to two different points.*

Since any metric space is Hausdorff (Exercise 3.5.10), the previous result implies

COROLLARY. *In a metric space, if a sequence converges, then its limit is unique.*

We now give a fundamental and very practical result in metric spaces:

THEOREM. *Let (X, d) be a metric space. Let $A \subset X$ be non-empty. Then we have:*

(1)

$$x \in \overline{A} \iff \exists x_n \in A : \ x_n \longrightarrow x.$$

(2)

$$A \text{ is closed} \iff \forall x_n \in A : (x_n \longrightarrow x \implies x \in A).$$

EXAMPLES.

(1) *In standard* $\mathbb{R}$, $(0, 1]$ *is not closed.*
(2) *In standard* $\mathbb{R}$ *again,* $\{\frac{1}{n} : n \in \mathbb{N}\}$ *is not closed.*

REMARK. Doubtlessly, the second result in the previous theorem can also be used to show that a given set is (or is not) open. See Exercises 4.3.21 & 4.3.22.

4.1.3. Sequential Continuity. We now come to a practical definition of continuity. But, in a general setting, it not as strong as the definition of continuity seen in the beginning of this chapter.

DEFINITION. *Let X and Y be two topological space and let $f : X \to Y$ be a function. Then we say that f is **sequentially continuous** at $x \in X$ if for any convergent sequence (x_n) to x, $(f(x_n))$ converges to $f(x)$.*

As usual, if f is sequentially continuous at each $x \in X$, then we say that f is sequentially continuous on X.

REMARK. Continuity implies sequential continuity but not vice versa. See Exercise 4.3.16. However, the two notions match in a metric space and we have

PROPOSITION (For a proof see Exercise 4.3.16). *Let (X, d) and (Y, d') be two metric spaces and let $f : (X, d) \to (Y, d')$ be a function. Then f is continuous on X iff it is sequentially continuous on X.*

4.2. True or False: Questions

QUESTIONS. Comment on the following questions/statements and indicate those which are false and those which are true when this applies. Justify your answers.

(1) The identity function between two topological spaces is always continuous.
(2) Let X and Y be two topological spaces. If f is a continuous function from X into Y, then for each open set U in X, $f(U)$ is open in Y.
(3) Let X and Y be two topological spaces. Assume that $f : X \to Y$ is some function and let f_A be its restriction to $A \subset X$. Then

$$f \text{ is continuous} \iff f_A \text{ is continuous.}$$

(4) Let X be a topological space and let $A \subset X$. Then

$$x \in \overline{A} \Longleftrightarrow \exists x_n \in A : \ x_n \longrightarrow x.$$

(5) The sequence $\left(\frac{1}{n}\right)$ converges to zero.

(6) Criticize the following proof: In the co-finite topology on $\mathbb{R}$, the sequence $\frac{1}{n}$ converges to any point in $\mathbb{R}$. To prove it, assume that $\frac{1}{n} \not\to x$ for all $x \in \mathbb{R}$. This means that

$$\exists U \in \mathcal{V}(x), \forall N \in \mathbb{N} \ \exists n \ (n \geq N \text{ and } \frac{1}{n} \notin U).$$

Hence $\frac{1}{n} \in U^c$ for all but finitely many n and this is a contradiction since U^c is finite!

(7) Find fault with the following reasoning: In usual $\mathbb{R}$, consider the set

$$A = \{x_n = n^2 : \ n \in \mathbb{N}\}.$$

Then A is not closed as (x_n) does not converge and hence it cannot have a limit belonging to A.

(8) In the usual topology, let $f : \mathbb{R}^*_+ \to \mathbb{R}$ be such that $f(x) = \ln x$. Then f is continuous, but $[0, 1]$ is closed in $\mathbb{R}$ and yet its preimage is not closed in $\mathbb{R}$. Is there anything wrong with that?

(9) If f is continuous and $-f$ is well-defined, then $-f$ is continuous.

(10) Let X be a topological space and let (x_n) be a convergent sequence in X. Then

$$X \text{ is Hausdorff } \Longleftrightarrow (x_n) \text{ has a unique limit.}$$

(11) Let X be a topological space. Endow $\mathbb{R}$ with its usual topology. Let $f : X \to \mathbb{R}$ be some function. If we come to show that $f^{-1}((a, b))$ (a and b real numbers with $a < b$) is open in X, then f is continuous.

(12) When are two sets homeomorphic? When are they not homeomorphic?

(13) Let X and Y be two topological spaces and let f be a continuous mapping from X into Y. Assume that $A \subset X$ is dense. Then $f(A)$ is dense in $f(X)$.

(14) Separability is a topological property.

(15) A bijective mapping between two topological spaces is continuous.

(16) An open, closed and bijective mapping between two topological spaces is continuous.

(17) Let X be a topological space and let $A \subset X$. Let $f : X \to X$ be a bijective mapping. Then

$$f \text{ is a homeomorphism iff } f(\overline{A}) = \overline{f(A)}.$$

(18) Every continuous bijection is a homeomorphism.

(19) Every continuous and open bijection is a homeomorphism.

(20) Hausdorffness is a topological property.

(21) Let X and Y be two topological spaces and let $f : X \to Y$ be a continuous function. Let $A \subset X$. If $x \in A'$, then $f(x) \in (f(A))'$.

(22) Let f be a map between two topological spaces X and Y such that Y is also assumed to be separated. Let $G_f = \{(x, f(x)) : x \in X\} \subset X \times Y$ be the graph of f. Then

$$f \text{ is continuous} \Longleftrightarrow G_f \text{ is closed}.$$

4.3. Exercises With Solutions

Exercise 4.3.1. Let X and Y be two topological spaces. Let $f : X \to Y$ be a function. If f continuous in the following cases?

(1) $f(x) = x$, X is the indiscrete topology and $Y = X$ equipped with the discrete topology;

(2) $f(x) = e^x$, $X = (\mathbb{R}, |\cdot|)$ and $Y = \mathbb{R}$ endowed with the discrete topology.

(3) $f(x) = x^2$, $X = (\mathbb{R}, T)$ and $Y = (\mathbb{R}, |\cdot|)$ where

$$T = \{\varnothing, \mathbb{R}\} \cup \{(a, +\infty), \ a \in \mathbb{R}\};$$

(4) f arbitrary, X discrete and Y arbitrary;

(5) f is the constant function;

(6) f is the canonical injection.

Exercise 4.3.2. Let $X = \{a, b, c\}$ and let $T = \{\varnothing, \{a\}, \{b\}, \{a, b\}\}$ be a topology on X. Let $f : X \to X$ be the function defined by

$$f(a) = a, \ f(b) = c \text{ and } f(c) = b.$$

Is f continuous at a? at b? at c?

Exercise 4.3.3. Let X and Y be two topological spaces. Let $f : X \to Y$ be a function. Show that

$$f \text{ is continuous} \Longleftrightarrow \forall U \subset Y : f^{-1}(\overset{\circ}{U}) \subset \overbrace{f^{-1}(U)}^{\circ}.$$

Exercise 4.3.4. Let X be an uncountable set. We endow X with the co-finite topology (see Exercise 3.3.22) and denote it by T. We also equip X with the "co-countable topology" (see Exercise 3.3.24) and we denote it by T'. Let $f : T' \to T$ be defined for any $x \in X$ by $f(x) = x$. Is f a homeomorphism?

Exercise 4.3.5. Let (X, d) and (X, d') be two metric spaces, where d' is defined for all $x, y \in X$ by

$$d'(x, y) = \frac{d(x, y)}{1 + d(x, y)}.$$

Let $f : (X, d') \to (X, d)$ be defined by $f(x) = x$ for all $x \in X$.
  (1) Show that f is a homeomorphism.
  (2) Deduce that boundedness is not a topological property.

Exercise 4.3.6. Show that Hausdorffness is a topological property.

Exercise 4.3.7.
  (1) Show that:
      (a) Any two open intervals in $\mathbb{R}$ are homeomorphic.
      (b) Show that that $\mathbb{R}$ is homeomorphic to $(-1, 1)$. Is $\mathbb{R}$ homeomorphic to any open interval in $\mathbb{R}$?
  (2) Is $\overline{\mathbb{R}}$ (the extended real line) homeomorphic to $[-1, 1]$?

Exercise 4.3.8. Let f be a continuous function from a topological space X into $\mathbb{R}$. Let a be a real number and set

$$A = \{x \in X : f(x) = a\}.$$

Verify that A is closed in X.

Exercise 4.3.9. Show that following sets are closed in X
  (1) $A = \{(x, y) \in \mathbb{R}^2 : xy = 1\}$, $X = \mathbb{R}^2$;
  (2) $B = \{(x, y) \in \mathbb{R}^2 : x^2 + y^2 \leq 1\}$, $X = \mathbb{R}^2$;
  (3) (A bit of linear algebra) $C = \{A \in \mathcal{M}_n(\mathbb{R}) : \det A = 0\}$, $X = \mathcal{M}_n(\mathbb{R})$, the (vector) space of square matrices of order n with real entries and we associate with it any metric.

Exercise 4.3.10. Let $a \in \mathbb{R}$. Let $f : \mathbb{R} \to \mathbb{R}$ be a continuous where $\mathbb{R}$ is endowed with the usual topology. Set

$$A = \{x \in \mathbb{R} : f(x) \leq a\}.$$

  (1) Show that A is closed in $\mathbb{R}$.
  (2) Is the converse always true?

Exercise 4.3.11. Let T be a topology on X. Let $\mathbb{R}$ be endowed with its usual topology. Let $f : X \to \mathbb{R}$ be a function and let $a \in \mathbb{R}$. Show that

f is continuous iff $f^{-1}((a, +\infty))$ and $f^{-1}((-\infty, a))$ are open in T.

Exercise 4.3.12 (Different limits of $\left(\frac{1}{n}\right)$). Let $\left(\frac{1}{n}\right)_{n\geq 1}$ be a sequence in $\mathbb{R}$. To what point (s) does $\left(\frac{1}{n}\right)_{n\geq 1}$ converge to (if there is any) with respect to:

(1) the usual topology;
(2) the co-finite topology;
(3) the discrete topology;
(4) the indiscrete topology.

Exercise 4.3.13. Set $x_n = \frac{1}{n}$ for all $n \in \mathbb{N}$, and let

$$K = \left\{\frac{1}{n} : n \in \mathbb{N}\right\}.$$

(1) Prove that (x_n) does not converge to any point of $\mathbb{R}$ in the K-topology $\mathbb{R}_K$ (defined in Exercise 3.3.30). What about $(-x_n)$?
(2) Show that $-K$ is not closed in $\mathbb{R}_K$.
(3) Show that $(x_n) \to 0$ in $\mathbb{R}_\ell$. Can (x_n) have another limit in $\mathbb{R}_\ell$?
(4) Is $f : \mathbb{R}_K \to \mathbb{R}_K$, defined by $f(x) = -x$, continuous?
(5) Is $f : \mathbb{R}_\ell \to \mathbb{R}_\ell$, defined by $f(x) = -x$, continuous?

Exercise 4.3.14. Let X be an infinite countable set endowed with the co-countable topology (see Exercise 3.3.24).

(1) Show that the only convergent sequences are the eventually constant ones.
(2) Set $X = [0,3]$ and $A = [2,3]$.
  (a) Show that $1 \in A'$.
  (b) Let (x_n) be a sequence in A. Can (x_n) converge to 1?
  (c) What is then the conclusion?

Exercise 4.3.15. Let (X, d) be a metric space. Let (x_n) and (y_n) be two convergent sequences to x and y respectively.

(1) Show that $d(x_n, y_n) \to d(x, y)$.
(2) How can this result be interpreted?

Exercise 4.3.16 (Continuity Vs. Sequential Continuity). Let X and Y be two topological spaces and let $f : X \to Y$ be a function.

(1) Show that if f is continuous, then it is sequentially continuous.
(2) Give an example that shows that the converse is not always true.
(3) If X and Y are metric spaces, then show that f is continuous iff it is sequentially continuous.

Exercise 4.3.17. Prove that every *continuous* and *monotonic* mapping from $\mathbb{R}$ into $\mathbb{R}$ is open.

Exercise 4.3.18. Let $P : \mathbb{R} \to \mathbb{R}$ be a polynomial. Show that P is a closed mapping (hint: if (x_n) is a sequence and $P(x_n)$ is bounded, then is (x_n) bounded?).

Exercise 4.3.19. Let X and Y be two topological spaces and let $f : X \to Y$ be a continuous function. Assume that $A \subset X$. Show that f is continuous on A.

Exercise 4.3.20. Using sequences, are the following sets closed in X

(1) $A = (0, 1]$, $X = \mathbb{R}$;
(2) $B = \left\{ \frac{1}{n} : n \geq 1 \right\}$, $X = \mathbb{R}$;
(3) $C = \{(x, y) \in \mathbb{R}^2 : x^2 + y^2 < 1\}$, $X = \mathbb{R}^2$;
(4) $D = \{(x, y) \in \mathbb{R}^2 : 1 < x^2 + y^2 \leq 3\}$, $X = \mathbb{R}^2$?

Exercise 4.3.21. Using sequences, show that the set
$$A = \{(x, y) \in \mathbb{R}^2 : 0 < y < 1, xy = 1\}$$
is neither open nor closed in $\mathbb{R}^2$ (endowed with the euclidian metric).

Exercise 4.3.22. Is the set $A = \left\{ \frac{1}{n} : n \geq 1 \right\}$ open in $\mathbb{R}$?

Exercise 4.3.23. Let f and g be two continuous functions defined on a metric space X into another metric space Y.

(1) Let $A = \{x \in X : f(x) = g(x)\}$. Show that A is closed in X.
(2) Let B be a dense subset in X. If f and g coincide on B, then show that they do coincide on X.

Exercise 4.3.24. Check that the following function is not continuous at $(0, 0)$
$$f(x, y) = \begin{cases} \frac{xy}{x^2+y^2}, & (x, y) \neq (0, 0), \\ 0, & (x, y) = (0, 0). \end{cases}$$

Exercise 4.3.25. Let $f : X \to Y$ be a continuous and one-to-one mapping where X and Y are two topological spaces.

(1) Show that if Y is Hausdorff, then so is X.
(2) Give a counterexample showing that the hypothesis of the continuity cannot merely be dropped.
(3) Give a counterexample showing that the hypothesis of the injectivity cannot be completely eliminated.

Exercise 4.3.26. Let (X, d) be a metric space. Let $A \neq \varnothing$ and $B \neq \varnothing$ be two closed sets such that $A \cap B = \varnothing$. Define a real-valued function f on X by
$$f(x) = \frac{d(x, A)}{d(x, A) + d(x, B)}.$$

(1) Show that if g is a real-valued continuous function defined on X, then its zero set is closed. *Recall that the zero set of g, denoted by $Z(g)$ as in [11], is defined as*

$$Z(g) = \{x \in X : g(x) = 0\}.$$

(2) Show that f is a continuous function on X.
(3) Find $f^{-1}(\{0\})$ and $f^{-1}(\{1\})$.
(4) Deduce a converse of the result of Question 1, that is, every closed set can be regarded as the zero of some continuous real-valued function.
(5) (cf. Exercise 4.5.10) Establish the existence of two *disjoint open* sets U and V such that $A \subset U$ and $B \subset V$.

REMARK. The property in Question 5 (plus the closedness of points!) is usually referred to as *normality* or *regularity* which is part of the separation axioms in topological spaces (the reader may easily check that normality is more powerful than Hausdorffness). Thus the previous exercise tells us that a *metric space* is necessarily normal. The other properties of normal spaces are not discussed in the present manuscript.

Exercise 4.3.27 (The moving bump). Let (f_n) be the sequence of functions defined as

$$f_n(x) = \begin{cases} 0, & 0 \le x \le \frac{1}{n+1}, \\ 2n(n+1)(x - \frac{1}{n+1}), & \frac{1}{n+1} < x \le \frac{2n+1}{2n(n+1)}, \\ -2n(n+1)(x - \frac{1}{n}), & \frac{2n+1}{2n(n+1)} < x \le \frac{1}{n}, \\ 0, & \frac{1}{n} < x \le 1. \end{cases}$$

(This sequence is usually called the "*moving bump*"). Then all f_ns are obviously continuous on $[0, 1]$. Show that for all $n, m \in \mathbb{N}$ with $n \ne m$:

$$d_\infty(f_n, f_m) = 1$$

(where d_∞ denotes the supremum metric).

REMARK. The importance of this sequence of functions will be illustrated in the coming chapters.

Exercise 4.3.28. Let $X = C([0, 1], \mathbb{R})$ and

$$A = \{f \in X : f(0) = 0\}.$$

Endow X with the metrics d and d' of Exercise 2.3.12. Show that $\overline{A} = A$ with respect to d' and $\overline{A} = X$ with respect to d.

Exercise 4.3.29. Let $X = C([0, 1], \mathbb{R})$ be endowed with the *supremum metric* and let $A_a = \{f \in X : f(a) = 0\}$ where $a \in [0, 1]$.

(1) Show that A is closed in X.
(2) Deduce that the set

$$B = \{f \in X : \ g = 0\},$$

where g is f restricted to $I \subset [0, 1]$, is closed in X.

Exercise 4.3.30. Let X and Y be two topological spaces.
(1) Show that the projections

$$p : X \times Y \to X$$
$$(x, y) \ \mapsto p(x, y) = x$$

and

$$q : X \times Y \to X$$
$$(x, y) \ \mapsto q(x, y) = y$$

are continuous but they are not closed.

   HINT. *You may consider the set* $A = \{(x, y) \in \mathbb{R}^2 : xy = 1\}$.

(2) Show that p and q are open mappings.

Exercise 4.3.31. Let X be a topological space. The **diagonal** of X is defined to be the set

$$\triangle = \{(x, x) : \ x \in X\}.$$

(1) Show that X is Hausdorff iff its diagonal is closed.
(2) Deduce from the preceding question another proof of the first question of Exercise 4.3.23 in a more general context, that is, $f, g : X \to Y$ are continuous, X is any topological space and Y is Hausdorff.

Exercise 4.3.32. Let A, X and Y be three topological spaces. Let $f : A \to X \times Y$ be a function defined by $f(x) = (g(x), h(x))$.

(1) Show that f is continuous if and only if g and h are so.
(2) Can we have a similar result for functions of the type $f : X \times Y \to A$, i.e. if f is continuous at any $x \in X$ and f is continuous at any $y \in Y$, then f is continuous on $X \times Y$?

Exercise 4.3.33. Let $f : X \to Y$ be a map between two topological spaces X and Y where Y is Hausdorff.

(1) Show that if f is continuous, then its graph, G_f, is closed.
(2) Can the graph of a non-continuous function be closed?

4.4. Tests

Test 23. Let $\mathbb{R} \to \mathbb{R}_\ell$ be the function defined by $f(x) = x^2$ where $\mathbb{R}$ is equipped with the discrete topology and $\mathbb{R}_\ell$ is the lower limit topology on $\mathbb{R}$. Is f a homeomorphism?

Test 24. Let $f : \mathbb{R} \to \mathbb{R}^*$ be a continuous function. Let

$$A = \{x \in \mathbb{R} : \ f(x) = 0\}.$$

Why is A closed in $\mathbb{R}$?

Test 25. Is the identity map continuous from $\mathbb{R}$ endowed with the co-finite topology into $\mathbb{R}$ endowed with the topology of Exercise 3.3.25? What about its inverse?

Test 26. Equip $\mathbb{R}$ with the topology of Exercise 3.3.25. To what point (if there is any) does the sequence defined by $x_n = \frac{1}{n}$ converge to?

Test 27. Does the sequence $(-\frac{1}{n})$ converge to 0 in $\mathbb{R}_\ell$?

Test 28. Let X be a topological space and $A \subset X$. Let $Y = \{0, 1\}$ be equipped with the topology $T = \{\varnothing, \{1\}, Y\}$. Define a function $f : X \to Y$ by

$$f(x) = \begin{cases} 1, & x \in A, \\ 0, & x \notin A. \end{cases}$$

Show that f is continuous on X iff A is open in X.

Test 29. Let X be an indiscrete topological space. When is $f : X \to \mathbb{R}$ ($\mathbb{R}$ equipped with the usual topology) continuous?

Test 30. In $\mathbb{R}^2$ endowed with the euclidian metric, let

$$A = \left\{\left(x, \frac{1}{x}\right) : \ x > 0\right\} \text{ and } B = \{(y, 0) : \ y \in \mathbb{R}\}.$$

(1) Show that A and B are closed in $\mathbb{R}^2$.
(2) Is $A + B$ closed?

Test 31. Let X be a topological space and let $\triangle$ be its diagonal. Show that X and $\triangle$ are homeomorphic.

Test 32. Is the "topological property" an equivalence relation?

Test 33. Show that $(0, 1)$ is homeomorphic to $(0, +\infty)$.

Test 34. Let $f : (X, T) \to (X, T)$ the identity map, i.e. $f(x) = x$. Show that f has a closed graph if and only if X is Hausdorff.

4.5. More Exercises

Exercise 4.5.1. Let X and Y be two topological spaces. Let $A, B \subset X$ be such that $\overline{A} = \overline{B}$. Let $f : X \to Y$ be a continuous function. Show that $\overline{f(A)} = \overline{f(B)}$.

Exercise 4.5.2. Let T and T' be two topologies on a set X. Show that the following statements are equivalent:

(1) $T \subset T'$;
(2) The identity map $id : (X, T') \to (X, T)$ is continuous;
(3) The identity map $id : (X, T) \to (X, T')$ is open;
(4) The identity map $id : (X, T) \to (X, T')$ is closed.

Exercise 4.5.3. Consider the topology of Exercise 3.3.27 and let $(x_n)_n$ be a sequence in $X = [0, 2)$ which is convergent in $\mathbb{R}$ to 1. Show that $(x_n)_n$ converges with respect to T to any point of $[1, 2)$.

Exercise 4.5.4. Using Exercise 4.3.30 and known results on the sum and scalar-multiplication of continuous functions, show that real-valued polynomials on $\mathbb{R}^n$ (and taking their values in $\mathbb{R}$) are continuous.

Exercise 4.5.5. If X and Y are two topological spaces, show that $X \times Y$ and $Y \times X$ are homeomorphic.

Exercise 4.5.6. Let $a \in \mathbb{R}$. Let $f : \mathbb{R} \to \mathbb{R}$ be a continuous where $\mathbb{R}$ is endowed with the usual topology. Set

$$A = \{x \in \mathbb{R} : \ f(x) > a\}.$$

(1) Show that A is open in $\mathbb{R}$.
(2) Is the converse always true?

Exercise 4.5.7. In the usual topology, define a real-valued function f on $\mathbb{R}$ by

$$f(x) = \begin{cases} 1, & x = 0, \\ 0, & x \in \mathbb{R} \setminus \mathbb{Q}, \\ \frac{1}{q}, & x \in \mathbb{Q} \end{cases}$$

where x in the last line is written as $\frac{p}{q}$, $p \in \mathbb{Z}, q \in \mathbb{N}$ and p and q are coprime. This function is called the **Riemann function**.

Show that f is continuous at *every irrational* and it is not continuous at *every rational* number.

Exercise 4.5.8. Let X and Y be two topological spaces. Let A and B be two closed sets in X such that $A \cup B = X$. Let $f : A \to Y$ and $g : B \to Y$ be two continuous functions. Assume that f and g coincide on $A \cap B$.

Define a function h as follows

$$h(x) = \begin{cases} f(x), & x \in A, \\ g(x), & x \in B. \end{cases}$$

Show that h is continuous.

REMARK. The result in the previous exercise is usually called the "**pasting lemma**".

Exercise 4.5.9. Show that the Cantor set does not have isolated points.

Exercise 4.5.10 (cf. Exercise 4.3.26). Let (X, d) be a metric space. Let A and B be two subsets of X such that **either** they are closed and disjoint **or** they satisfy $\overline{A} \cap B = A \cap \overline{B} = \varnothing$. Set

$$U = \{x \in X : d(x, A) < d(x, B)\} \text{ and } V = \{x \in X : d(x, B) < d(x, A)\}.$$

Show that U and V are two disjoint and open sets in (X, d) which contain A and B respectively.

Exercise 4.5.10 (cf. Exercise 4.3.17). Show that if $f : A \to \mathbb{R}$ is increasing and open $(A \subset \mathbb{R})$, then f is continuous.

Exercise 4.5.11. Prove that two *discrete* spaces are homeomorphic iff they have the same cardinality.

Exercise 4.5.12. Show that a circle deprived of one point is homeomorphic to a straight line.

Exercise 4.5.13. Let X and Y be two topological spaces, and let $f : X \to Y$ be a continuous function. Show that the graph of f is homeomorphic to X.

Exercise 4.5.14. We endow $\mathbb{R}$ with its usual topology. Let

$$A = \{-p : p \geq 2\} \text{ and } B = \left\{ n + \frac{1}{n} : n \geq 2 \right\}.$$

(1) Show that A and B are two closed sets in $\mathbb{R}$.
(2) Is $A + B$ closed in $\mathbb{R}$?

Exercise 4.5.16 (see [5] and the references therein). Let $f, g : \mathbb{R} \to \mathbb{R}$ be two functions.

(1) Can f and g have closed graph but $f + g$ does not?
(2) Show that if f and g are *nonnegative* and have a closed graph, then $f + g$ has a closed graph.

Exercise 4.5.17 ([8]). Let X and Y be two topological spaces. Let $f : X \to Y$ be any function. Denote its graph by G_f.

(1) Show that if f is an open surjection such that G_f is closed, then Y is Hausdorff.
(2) Show that if f is one-to-one and continuous with G_f closed, then X is Hausdorff.
(3) Deduce that if f is a homeomorphism of X onto Y, then both X and Y are Hausdorff.

Exercise 4.5.15. Let X be a topological space and let $\mathcal{R}$ be an equivalence relation on X. Define

$$G_{\mathcal{R}} = \{(x, y) \in X^2 : x\mathcal{R}y\}$$

(called the graph of $\mathcal{R}$).

(1) Show that if $X/\mathcal{R}$ is Hausdorff, then $G_{\mathcal{R}}$ is closed.
(2) Is the converse always true? Justify your answer.
(3) Assume now that $G_{\mathcal{R}}$ is closed and that the quotient map is open. Show that $X/\mathcal{R}$ is Hausdorff.

Exercise 4.5.16. A **direct set** is a couple $(I, \prec)$ where I is a set and $\prec$ is a relation which is: reflexive, transitive and such that: For any $a, b \in I$, there is $c \in A$: $a \prec c$ and $b \prec c$.

A **net** in a set X is a sequence $(x_i)_i \subset X$ indexed by a direct set $(I, \prec)$.

It is clear that this notion generalizes that of a sequence. We also use the definition of a convergent net as the one of a sequence (with the obvious changes). Other topological notions involving sequences may be defined in terms of a net too.

(1) Give an example of a direct set.
(2) Let X, Y be two topological spaces and let $A \subset X$ and $x \in A$. Prove that:
   (a) x lies in the closure of A iff x is a limit of a net $(x_i)_{i \in I}$ from A.
   (b) X is Hausdorff iff every convergent net has a unique limit.
   (c) A function $f : X \to Y$ is continuous at x iff for every net $(x_i)_{i \in I}$ converging to x, the net $(f(x_i))_{i \in I}$ converges to $f(x)$.

CHAPTER 5

Compact Spaces

5.1. What You Need to Know

5.1.1. Compactness: General Notions.

DEFINITION. *Let A and I be two sets. A collection $\{U_i\}_{i \in I}$ of sets is said to be a **cover** (or a **covering**) for A if*

$$A \subset \bigcup_{i \in I} U_i.$$

*If all U_i are open, then we say an **open cover**.*

EXAMPLES.
(1) *Let X be a topological space. Then $\{X\}$ is an open (and a closed) cover for any subset A of X.*
(2) *In usual $\mathbb{R}$, $\{[-n, n)\}_{n \in \mathbb{N}}$ is a cover for $\mathbb{R}$. Similarly, $\{(-n, n)\}_{n \in \mathbb{N}}$ is an open cover for $\mathbb{R}$.*

DEFINITION. *Let (X, T) be a topological space. Then X is called **compact** if every open cover of X has a finite subcollection (also called **subcover**) that still covers X.*

REMARK. In this book we do not impose the separation (i.e. the Hausdorffness) in the definition of compact sets. This is adopted by many authors. But, in some references, especially the French ones, they do add the separation in the definition of compact sets. One of the reasons, is to exclude spaces not too interesting in the compactness such as the indiscrete spaces. But this a dilemma as many interesting spaces will be excluded too such as $\mathbb{R}$ in the co-finite topology.

However, many applications arise in the setting of metric spaces where the two definitions coincide.

REMARK. We may say a compact space or a compact set (or a compact subspace or a compact subset), but we should not consider this as a source of polemic.

REMARK. Let X be a topological space and let $A \subset X$. Open sets in A are of the form $A \cap U$ where U is open in X. So if we want to prove that the subspace A is compact, we may just consider covers constituted of open sets in X (instead of open sets in A).

EXAMPLES.

(1) In usual $\mathbb{R}$, $A = [1, \infty)$ is not compact since $\{(0, n)\}_{n \in \mathbb{N}}$ is an open cover for A which is not reducible to a finite subcover for A as the union of elements of any finite subcover is of the form $(0, N)$ for some natural N.

(2) Every finite set (in any topological space) is compact. For a proof see Exercise 5.3.8.

Another (fundamental) example is the following (for a proof see Exercise 5.3.5):

THEOREM (Heine-Borel). Every closed and bounded interval $[a, b]$ in (usual) $\mathbb{R}$ is compact.

In point of fact, the previous extends to $\mathbb{R}^n$ (also due to Heine-Borel) and we have

THEOREM. In the usual topology, every bounded and closed space of $\mathbb{R}^n$ is compact.

The next two results are devoted to properties of compact spaces.

PROPOSITION. Let (X, d) be a metric space and let $A \subset X$ be compact. Then A is bounded.

THEOREM. Every closed subspace of a compact space is compact, and every compact subspace of a **Hausdorff** space is closed.

DEFINITION. Let X be a topological space and let A be a subspace of X. We say that A is **relatively compact** if $\overline{A}$ is compact.

EXAMPLES. In the usual topology, $(-1, 2)$ is relatively compact in $\mathbb{R}$ since $[-1, 2]$ is compact.

Now, we give a definition (more commonly known in $\mathbb{R}$ as the **Bolzano-Weierstrass property**) of a different concept of compactness in general topological spaces (see [**10**]):

DEFINITION. A topological space X is called **limit point compact** if every infinite subset of X has a limit point.

The next result tells us the relationship between compactness and limit point compactness:

PROPOSITION. Every compact space is limit point compact.

REMARK. The converse of the previous result is not always true. As a counterexample (borrowed from [**10**]): Let $Y = \{a, b\}$ and equip it with an indiscrete topology. Then $X = \mathbb{N} \times Y$ is limit point compact but not compact.

DEFINITION. *A topological space X is said to be **locally compact** at $x \in X$ if there exists a compact subspace A of X such that $U \subset A$ where U is a neighborhood U of x.*
We say that X is locally compact if it is so at each of its points.

EXAMPLES.

(1) *Obviously, every compact space is locally compact.*
(2) $\mathbb{R}$ *and, $\mathbb{R}^n$ in general, are locally compact with respect to their usual topologies.*
(3) *A discrete topological space is locally compact.*

Here are some properties of locally compact spaces:

PROPOSITION.

(1) *A closed subspace of a locally compact space is locally compact.*
(2) *An open subspace of a locally compact Hausdorff space is locally compact.*

5.1.2. Compactness and Continuity.

THEOREM. *Let $f : X \to Y$ be a continuous map between two topological spaces. If X is compact, then so is its image $f(X)$.*

COROLLARY. *Let (X, T) be a compact topological space and let (Y, d) be a metric space. If $f : X \to Y$ is continuous, then it is bounded.*

COROLLARY. *Let X be a compact topological space. If $f : X \to \mathbb{R}$ is continuous, then $\sup f(X)$ and $\inf f(X)$ exist (i.e. f attains its bounds on X).*

The following examples show that the hypotheses in the previous results cannot merely be dropped:

EXAMPLES. *Let $f : X \to Y$ be a function.*

(1) $X = Y = \mathbb{R}$, $f(x) = x$ *is not bounded on the non-compact (non-bounded) X.*
(2) $X = (0, 1]$ *and $Y = [1, \infty)$, $f(x) = \frac{1}{x}$ is not on the non-compact (non-closed) X.*
(3) $X = \mathbb{R}$ *and $Y = (-1, 1)$, $f(x) = \frac{x}{1+|x|}$ does not attain its bounds on the non-compact (non-bounded) X.*

(4) $X = Y = (0, 1)$, $f(x) = x$ *does not attain its bounds on the non-compact (non-closed)* X.

We have already defined uniform continuity in metric spaces. The next is an important result:

THEOREM (Heine). *Let (X, d) and (X, d') be two metric spaces such that (X, d) is compact. Let $f : (X, d) \to (X, d')$ be a continuous function. Then f is uniformly continuous.*

We finish with a fundamental result on products of compact spaces.

THEOREM (Tychonoff). *If X and Y are two compact spaces, then so is their product $X \times Y$.*

REMARK. By induction, the previous result can be easily generalized to a product of finitely many spaces. In fact, the result remains valid even for an arbitrary product. This latter generalization is actually the Tychonoff theorem.

The reader is asked in Exercise 5.3.16 to prove the converse of the previous theorem.

5.1.3. Sequential Compactness and Total Boundedness.

DEFINITION. *Let X be a metric space. Then X is said to be **sequentially compact** if every sequence (x_n) in X has a convergent subsequence in X.*

REMARK. A priori, there seems to be no direct link to the definition of compactness met above. The next result tells us that, in fact, compactness, limit point compactness and sequential compactness all coincide in a metric space:

THEOREM. *Let (X, d) be a metric space. Then the following are equivalent:*

(1) *X is compact;*
(2) *X is sequentially compact;*
(3) *X is limit point compact.*

The proof of the previous result necessitates the notion of total boundedness, so we recall it here together with some of its properties. We shall need it again in later chapters.

DEFINITION. *Let (X, d) be a metric space. Let $\varepsilon > 0$. An ε-**net** for X is a subset A of X verifying $X \subset \bigcup_{x \in A} B(x, \varepsilon)$.*

DEFINITION. *Let (X, d) be a metric space and let $A \subset X$. We say that A is **totally bounded** if A has an ε-net for all ε.*

REMARK. A bounded set is not necessarily totally bounded. For a counterexample, consider the following infinite set

$$A = \{e_1 = (1, 0, 0, \cdots), e_2 = (0, 1, 0, \cdots), e_3 = (0, 0, 1, \cdots), \cdots\}$$

which is obviously a subset of ℓ^2 (cf Exercise 7.3.9). Then A is bounded but not totally bounded.

Now we list some properties of totally bounded sets:

THEOREM.

(1) *A subset of a totally bounded set is totally bounded.*
(2) *A totally bounded set is bounded.*
(3) *The closure of a totally bounded set is totally bounded.*

The connection between compactness and total boundedness is elucidated in the next theorem.

THEOREM. *A sequentially compact set is totally bounded.*

We finish this subsection with a practical result on uncountability.

THEOREM (see [10]). *A non-empty compact Hausdorff space X without any isolated point is uncountable.*

COROLLARY. *Every closed interval in $\mathbb{R}$ is uncountable.*

REMARKS.

(1) Of course, the previous corollary is not concerned with the very particular closed interval of the form $[a, a]$ ($a \in \mathbb{R}$).
(2) The fact that the result is concerned with closed sets is not a weakness. Indeed, if, for example, we want to prove that $[-1, 1)$ is uncountable, we consider $[-1, 1]$ which is uncountable by the previous corollary. Then taking out a point of an uncountable set does not make it countable!
(3) The Baire's theorem (see Chapter 8) can also be used to give a short proof of the uncountability of $[a, b]$.

5.2. True or False: Questions

QUESTIONS. Comment on the following questions/statements and indicate those which are false and those which are true when this applies. Justify your answers.

(1) Let X be a non-empty set. Let $\{U_i\}_{i \in I}$ be a cover for X. Then $X = \cup_{i \in I} U_i$.

(2) We saw in the first example of covers in the "What you need to know" section that in any topological space X, $\{X\}$ is an open cover for X and any subset of X. It is also clearly a subcover for X. Hence every subset of X is compact.

(3) The fact that we use open covers to define compact spaces is purely conventional. We could adopt the same definition using closed covers.

(4) In every topological space X, the countable intersection of closed, non-empty and decreasing subsets in X is non-empty.

(5) Let (C_n) be a sequence of non-empty decreasing and closed sets in a metric space X. If $f : X \to X$ is continuous, then

$$f\left(\bigcap_{n\in\mathbb{N}} C_n\right) = \bigcap_{n\in\mathbb{N}} f(C_n).$$

(6) $\varnothing$ is compact.

(7) The interval $[a, b]$ is always compact.

(8) The closure of a compact subset is compact.

(9) Let X and Y be two topological spaces. Let $f : X \to Y$ be a continuous function and let $A \subset Y$ be compact. Then $f^{-1}(A)$ is compact.

(10) In a topological space, a subspace is compact if and only if it is closed and bounded.

(11) Let T and T' be two topologies on the same set X such that $T \subset T'$. If X is compact with respect to T', it will be so with respect to T. What about the converse?

(12) Criticize the following proof of the compactness of $\mathbb{R}$ with respect to the co-finite topology: Let $\{U_i\}_{i\in I}$ be an open cover of $\mathbb{R}$, i.e. $\mathbb{R} \subset \bigcup_{i\in I} U_i$. Now, assume $\mathbb{R} \not\subset \bigcup_{i=1}^{n} U_i$. Then

$$\mathbb{R} \subset \left(\bigcup_{i=1}^{n} U_i\right)^c = \bigcap_{i=1}^{n} U_i^c \text{ (a finite set)},$$

i.e. $\mathbb{R}$ would have to be finite and we arrived at a contradiction. Thus $\mathbb{R}$ is compact with respect to the co-finite topology.

(13) Let $f : [a, b] \to \mathbb{R}$ be a continuous function. Then f is bounded.

(14) $(0, 2)$ is relatively compact.

(15) The closed unit ball is always compact.

(16) Every compact set is closed.

(17) Compactness is a topological property.

(18) Sequential compactness is a topological property.

(19) Local compactness is preserved under continuous maps.
(20) Local compactness is a topological property.
(21) The union of two locally compact spaces remains locally compact.
(22) In a metric space, every finite part is totally bounded.
(23) Total boundedness is a topological property.

5.3. Exercises With Solutions

Exercise 5.3.1. Using (only) the definition of a compact set show that $\mathbb{R}$, $[0, +\infty)$ and $(0, 1)$ are not compact in $\mathbb{R}$ (in the usual topology).

Exercise 5.3.2. Indicate which of the following sets are compact in X

(1) $A = \mathbb{Q}$, $X = \mathbb{R}$;
(2) $A = \{\frac{1}{n} : n \in \mathbb{N}\}$, $X = \mathbb{R}$;
(3) $A = \mathbb{Q} \cap [0, 1]$, $X = \mathbb{R}$;
(4) $A = [a, b]$, $B = \mathbb{R}$, C an infinite set; X endowed with the discrete topology;
(5) $A = \{(x, y) \in \mathbb{R}^2 : x^2 + y^2 = 1\}$, $B = \{(x, y) \in \mathbb{R}^2 : x^2 + y^2 < 1\}$, $C = \{(x, y) \in \mathbb{R}^2 : x^2 + y^2 > 1\}$; $X = \mathbb{R}^2$;
(6) $A = \{(x, y) \in \mathbb{R}^2 : x \geq 1, \ 0 \leq y \leq \frac{1}{x}\}$; $X = \mathbb{R}^2$;
(7) $A = \{(x, \frac{1}{x}) : 0 < x \leq 1\}$, $B = \{(x, \sin\frac{1}{x}) : 0 < x \leq 1\}$; $X = \mathbb{R}^2$;
(8) $A = \bigcap_{n \in \mathbb{N}} B\left(0_{\mathbb{R}^2}, 1 + \frac{1}{n}\right)$; $X = \mathbb{R}^2$?

   (The topology of $\mathbb{R}^2$ in Questions 5 to 8 is the standard one).

Exercise 5.3.3.

(1) Show that the following set is not compact in $\mathbb{R}^2$ with respect to the standard topology
$$A = \{(x, y) \in \mathbb{R}^2 : \ x + y^3 = 1\}.$$

(2) What about the set
$$B = \{(x, y) \in \mathbb{R}^2 : x^4 + y^2 = 1\}?$$

Exercise 5.3.4. In the usual topology of $\mathbb{R}$, show that
$$A = \left\{\frac{1}{n} : \ n \in \mathbb{N}\right\}$$
is not compact using open covers.

Exercise 5.3.5. Show that the closed bounded interval $[a, b]$ (a and b are reals with $a \leq b$) is compact in usual $\mathbb{R}$.

REMARK. Another proof of the compactness of $[a, b]$ may be found in Exercise 7.3.23.

Exercise 5.3.6. Consider $\mathbb{R}_K$, the K-topology on $\mathbb{R}$.
  (1) Is $[0, 1]$ compact in $\mathbb{R}_K$?
  (2) What about any set that contains $K = \{\frac{1}{n} : n \in \mathbb{N}\}$?

Exercise 5.3.7. Let X be a topological space which is Hausdorff. Let $(x_n)_n$ a sequence in X which converges to a. Show that the set $A = \{x_n : n \in \mathbb{N}\} \cup \{a\}$ is compact.

Exercise 5.3.8. Let X be a topological space.
  (1) Show that every finite set in X is compact.
  (2) Deduce that if X is given the *discrete* topology, then a set is compact if and only if it is finite.

Exercise 5.3.9.
  (1) Show that the union of two (and hence of a finite number) of compact sets is compact. Is this true for an arbitrary union?
  (2) Show that the arbitrary intersection of compact sets in a Hausdorff topological space is always compact.

Exercise 5.3.10. Let (X, d) be a metric space and let $A \subset X$. We know from earlier chapters that the union $\cup_{x \in A} B_c(x, r)$ need not be closed, where $B_c(x, r)$ is the closed ball of center x and radius $r > 0$.

Show that if A is compact, then for any $r > 0$, $\cup_{x \in A} B_c(x, r)$ is closed in X (hint: you may show that

$$\bigcup_{x \in A} B_c(x, r) = \{t \in X : \inf_{x \in A} d(x, t) \leq r\}).$$

Exercise 5.3.11. Is $\mathbb{R}$ compact with respect to $X = \mathbb{R}$ endowed with the co-finite Topology?

Exercise 5.3.12. Consider $\mathbb{R}$ with respect the co-countable topology (see Exercise 3.3.24).
  (1) Is $\mathbb{R}$ compact?
  (2) Is $[0, 1]$ compact?
  (3) Let A be a countable set. Is A compact?

Exercise 5.3.13. On $\mathbb{R}$, we define

$$T = \{\varnothing\} \cup \{U \subset \mathbb{R} : U^c \text{ is compact in the usual } \mathbb{R}\}.$$

  (1) Show that T is a topological space on $\mathbb{R}$.

(2) Can T be Hausdorff?

(3) Show that $\mathbb{R}$ is separable with respect to T.

(4) Show that $\mathbb{R}$ is compact with respect to T.

Exercise 5.3.14. By going back to Exercise 3.3.26 and taking $a = 1$, show that $[-1, 1]$ is compact in T.

Exercise 5.3.15. In the usual topology, explain why $(0, 1)$ is not compact using limit point compactness.

Exercise 5.3.16. Let X and Y be two topological spaces. Show that if $X \times Y$ is compact, then X and Y are also compact.

Exercise 5.3.17. Let $\mathbb{Q}$ be the metric space associated with the metric d defined by $d(x, y) = |x - y|$.

(1) Show that the set $A = \{x \in \mathbb{Q} : 2 < x^2 < 3\}$ is closed and bounded but not compact.

(2) Is A open in $\mathbb{Q}$?

Exercise 5.3.18. Consider the following metric defined on $\mathbb{R}$ by $d(a, b) = \inf\{|b - a|, 1\}$.

(1) Is $(\mathbb{R}, d)$ bounded?

(2) Using the sequence defined by $a_n = n$ or otherwise, show that $(\mathbb{R}, d)$ is not compact with respect to this metric.

(3) What can you deduce from the previous question?

Exercise 5.3.19.

(1) Show that $\mathbb{R}$ is not sequentially compact with respect to the metric δ of Exercise 2.3.26.

(2) What can deduce from the previous question?

Exercise 5.3.20. Let $X = C([0, 1], \mathbb{R})$ be equipped with the supremum metric.

(1) What is $\dim X$?

(2) Show that unit closed ball is not sequentially compact in X.

(3) Can the unit closed ball be compact in X?

(4) Is X locally compact?

Exercise 5.3.21. We endow $\mathbb{R}$ with two topologies, one is the cofinite one and we denote it by X, the other is the discrete one and we denote it by Y. Let $f : X \to Y$ be a function defined for all $x \in \mathbb{R}$ by $f(x) = x^2$.

(1) Is $f(\mathbb{R})$ compact?

(2) What can you deduce from the previous question?

Exercise 5.3.22. In the usual topology, is $[0, 1]$ homeomorphic to $[0, \infty)$? to $(0, 1]$?

Exercise 5.3.23. Let X be a compact *metric* space. Let $f : X \to X$ be continuous. If (A_n) is a sequence of decreasing nonvoid and closed sets in X, then show that

$$f\left(\bigcap_{n \in \mathbb{N}} A_n\right) = \bigcap_{n \in \mathbb{N}} f(A_n).$$

Exercise 5.3.24.

(1) Indicate among the following spaces X those which are locally compact and those which are not:
   (a) X is a compact topological space,
   (b) $X = \mathbb{R}$ in the usual topology,
   (c) $X = \mathbb{Q}$ in the usual topology,
   (d) $X = \mathbb{R} \setminus \mathbb{Q}$ in the usual topology,
   (e) X is a discrete topological space,
   (f) X equipped with the topology of Exercise 3.3.25.
(2) Give an example of a Hausdorff space which is not locally compact and one which is locally compact.

Exercise 5.3.25. Give an example that shows the continuous image of a locally compact space need not be locally compact.

Exercise 5.3.26. Let X be a locally compact space.

(1) Show that every closed subspace in X is locally compact.
(2) Show that every open subspace in X is locally compact provided X is *Hausdorff*.
(3) In usual $\mathbb{R}^2$, say why $\{(0, 0)\}$ and $\{(x, y) \in \mathbb{R}^2 : x > 0\}$ are locally compact.
(4) In the usual topology again, is

$$\{(0, 0)\} \cup \{(x, y) \in \mathbb{R}^2 : x > 0\}$$

locally compact?
(5) What is then the conclusion?

Hint: For the second question, use the following result: Let X be a locally compact Hausdorff space. Let $x \in X$ and U be a neighborhood of x. Then there is a neighborhood of x, denoted by V such that its closure is compact and it is included in U.

Exercise 5.3.27. Let (X, d) be a *compact* metric space. Let $f : X \to X$ be a function satisfying

$$d(f(x), f(y)) < d(x, y), \ \forall x \neq y.$$

(1) Can we say that there exists a $k \in [0,1)$ such that
$$d(f(x), f(y)) \le kd(x,y), \ \forall x, y.$$

(2) Show that f has a unique fixed point, that is, there is one and only one point $x \in X$ such that $f(x) = x$.

Exercise 5.3.28. Let (X, d) be a *compact* metric space. Prove that (X, d) is separable.

Exercise 5.3.29. When is (X, d) totally bounded if X is an arbitrary set and d is the discrete metric?

5.4. Tests

Test 35. Using open covers, show that $A = (0,1] \cup \{2\}$ is not compact w.r.t. usual $\mathbb{R}$.

Test 36. In the topology of Exercise 3.3.25, is $\{a\}$ compact? What about X? What can you deduce from that?

Test 37. Let (X, T) be separated topological space. Let (X, S) be a compact topological space. Show that if $T \subset S$, then $T = S$.

Test 38. Show that $\mathbb{N}$ is not compact with respect to the topology of Exercise 3.5.7.

Test 39. Is $\mathbb{R}$ compact in $\mathbb{R}_\ell$?

Test 40. Is $\mathbb{Q}$ relatively compact in $\mathbb{R}$ with respect to the co-finite topology?

Test 41. Is the unit sphere in $\mathbb{R}^3$ homeomorphic to $\mathbb{R}^2$?

Test 42. Let X be a topological space. Let A be a finite subset of X. Show that A is sequentially compact.

Test 43. Does the quotient of a compact space remain compact?

Test 44. Define a function f on $[0,1]$ by
$$f(x) = \begin{cases} 0, & x = 0, \\ x \ln x, & 0 < x \le 1. \end{cases}$$

Is f uniformly continuous on $[0,1]$?

Test 45. Let $A_i = \{1, -1\}$. Without caring much about the (arbitrary) product topology involved, is $\prod_{i \in I} A_i$ compact (I is a set)?

5.5. More Exercises

Exercise 5.5.1. Consider the topology of Exercise 3.3.27. Is $[0, 2)$ compact in this topology?

Exercise 5.5.2. Show that the set
$$A = \{(x, y) \in \mathbb{R}^2 : x^2 - xy + y^2 \leq 1\}$$
is compact in $\mathbb{R}^2$.

Exercise 5.5.3. Let $\mathbb{N}$ be endowed with the co-finite topology. Show that the set of even integers is compact in this topology. Is it closed?

Exercise 5.5.4. Show that $\mathbb{N}$ endowed with the co-countable topology is not compact.

Exercise 5.5.5. Find an example of two compact subspaces A and B (in a non-Hausdorff space) such that $A \cap B$ is not compact.

Exercise 5.5.6. Let T and T' be two topologies on the same set X which are assumed to be compact and Hausdorff. Prove that T and T' are either equal or not comparable.

Exercise 5.5.7. Show that each compact metric space is separable.

Exercise 5.5.8. Show that the Cantor set is compact. Show also that it is uncountable.

Exercise 5.5.9. Let X and Y be two *separated* topological and *compact* spaces. Let G_f be the graph of f. Show that f is continuous if and only if G_f is closed in $X \times Y$.

Exercise 5.5.10. ([5]) Let $f : \mathbb{R} \to \mathbb{R}$ be a function having a closed graph. Show that if K is a compact set in $\mathbb{R}$, then $f^{-1}(K)$ is closed in $\mathbb{R}$.

Exercise 5.5.11. Show that $C([0, 1], \mathbb{R})$ (with respect to the supremum metric) is not compact using open covers.

Exercise 5.5.12. (Hausdorff metric) Let (X, d) be a compact metric space. Denote the collection of closed sets in (X, d) by $Cl(X)$. Define a function d on $Cl(X) \times Cl(X)$ by
$$d_H(A, B) = \max(\sup_{a \in A} d(a, B), \sup_{b \in B} d(b, A))$$
where $d(x, C) = \inf_{c \in C} d(x, c)$.

Show that d_H is a metric on $Cl(X)$ (d_H is called the **Hausdorff metric**).

REMARK. It is important to consider closed sets in X which they then become compact. If the compactness hypothesis is dropped, then d_H need not remain a metric anymore. For instance, the Hausdorff "metric" applied to the sets $\{1\}$ and $(-\infty, 1]$ is infinite, hence it is necessary to have bounded sets. It is also necessary to have closed sets (take $A = [-1, 1]$ and $B = (-1, 1)$).

Exercise 5.5.13. We know that "*a non-empty compact Hausdorff space X without any isolated point is uncountable*". Give an example showing that the hypothesis *Hausdorff* cannot be dispensed with.

Exercise 5.5.14. (cf. Exercise 4.3.26) Show that every compact Hausdorff space is normal.

Exercise 5.5.15. Show that local compactness is a topological property.

Exercise 5.5.16. Let (X, d) be a *compact* metric space. Let f be an isometry from X *into* X. Prove that f is onto (hint: show that $X \subset f(X)$).

Exercise 5.5.17. Let X be a locally compact topological space. Set $\tilde{X} = X \cup \{\infty\}$ where ∞ is something which does not belong to X. Then $\tilde{X}$ is called the **Alexandroff one-point compactification** of X. Then a topology can be defined on $\tilde{X}$ by declaring a set open iff it is either open in X or it is of the form $\tilde{X} \setminus K$ where K is compact in X. This will be the topology associated with $\tilde{X}$ by default.

A known result then says that X is homeomorphic to $\tilde{X} \setminus \{\infty\}$.

(1) Prove that $\tilde{X}$ is compact.
(2) Verify that X is closed in $\tilde{X}$ iff it is compact.
(3) Deduce that if X is not compact, then it is dense in $\tilde{X}$.
(4) Show that if Y and Z are two compact spaces, then a homeomorphism $f : Y \setminus \{y\} \to Z \setminus \{z\}$ (where $y \in Y$ and $z \in Z$) can be extended to a homeomorphism $g : Y \to Z$ by setting $g(y) = z$.
(5) Show that in the usual topology, the Alexandroff one-point compactification of $(-1, 0]$ is $[-1, 0]$ and that of $\mathbb{R}$ is homeomorphic to the unit circle in $\mathbb{R}^2$.

Exercise 5.5.18. Show that bounded subsets of $\mathbb{R}^n$ are totally bounded.

Exercise 5.5.19. We first give the following definition:

DEFINITION. *Let X be a metric space and let $A \subset X$. Let $\mathcal{U} = \{U_i\}_{i \in I}$ be an open cover for A. A real number $\varepsilon > 0$ is said to a*

Lebesgue number for $\mathcal{U}$ if for any $x \in A$, there is some set U $(U \ni x)$ in $\mathcal{U}$ verifying $B(x, \varepsilon) \subset U$.

Prove that any open cover of a sequentially compact metric space has a Lebesgue number.

CHAPTER 6

Connected Spaces

6.1. What You Need to Know

6.1.1. Connectedness.

DEFINITION. *A topological space X is said to be **connected** if the only closed and open sets (that is, the only clopen sets) in X are $\varnothing$ and X itself.*

EXAMPLES.

(1) *Evidently, every indiscrete topological space is connected.*
(2) *A discrete topological space X is never connected unless $\operatorname{card} X = 1$ (in which case the discrete and indiscrete spaces coincide!).*

There exist equivalent definitions of connectedness. They are gathered in the next theorem:

THEOREM. *Let X be a topological space. Then X is connected iff one of the following occurs:*

(1) *Any continuous function from X onto the discrete space $\{0, 1\}$ is constant.*
(2) *X does not admit any open partition.*

REMARK. By an **open partition**, we mean a couple of non-empty open subspaces A and B in X such that $A \cap B = \varnothing$ and $A \cup B = X$.

REMARK. Notice that there is nothing special about the set $\{0, 1\}$, any $\{a, b\}$ will do!

Before giving the characterization of connected subspaces of the usual real line $\mathbb{R}$, recall the following:

DEFINITION. *A non-void subset A of $\mathbb{R}$ is called an **interval** if:*

$$\forall x, y \in A, \forall z \in \mathbb{R} : (x < z < y \implies z \in A).$$

EXAMPLES.

(1) *$\mathbb{Q}$ is not an interval.*
(2) *$(0, 1) \cup \{2\}$ is not an interval.*

The following result characterizes connected subspaces of the usual real line.

THEOREM. *Every interval is connected. Conversely, the only connected subspaces of* $\mathbb{R}$ *are the intervals.*

The next result is fundamental.

THEOREM. *Let X and Y be two topological spaces. Let $f : X \to Y$ be continuous. If X is connected, then $f(X)$ is connected too.*

COROLLARY. *Connectedness is a topological property.*

COROLLARY. *Let X and Y be two topological spaces. Let $f : X \to Y$ be continuous. If X is connected, then G_f, i.e. the graph of f, is connected.*

COROLLARY. *Let X be a topological space which is connected. If $f : X \to \mathbb{R}$ is continuous, then $f(X)$ is an interval.*

REMARK. The previous corollary generalizes the well-known intermediate value theorem, where setting $X = [a, b]$ takes us back to it.

We finish this subsection with two results on unions of connected spaces. The last of the two is also important for the ensuing subsection.

PROPOSITION. *If X and Y are two non-empty connected sets such that $\overline{X} \cap Y \neq \varnothing$, then $X \cup Y$ is connected.*

REMARK. The condition $\overline{X} \cap Y \neq \varnothing$ in the preceding proposition cannot be weakened to $\overline{X} \cap \overline{Y} \neq \varnothing$. For a counterexample see Question 4 of Exercise 6.3.1.

PROPOSITION. *Let X be a topological space and let $(A_i)_{i \in I}$ be collection of connected subspaces of X such that $\cap_{i \in I} A_i \neq \varnothing$. Then $\cup_{i \in I} A_i$ is connected.*

6.1.2. Components.

THEOREM. *Let X be a topological space and let $x \in X$. Then there exists a maximal or largest (with respect to "$\subset$") connected subspace of X, denoted by C_x, containing x. Moreover, C_x is closed in X.*

DEFINITION. *Let X be a topological space, and let $x \in X$. The maximal, closed and connected subspace C_x containing x is called a (connected)* **component** *of X.*

REMARK. We may easily define an equivalence relation on X as: $x \mathcal{R} y \iff y \in C_x$. Hence, the components of X constitute a partition of X.

EXAMPLE. $\mathbb{R}\backslash\{1\}$ *has two components: specifically* $(-\infty, 1)$ *and* $(1, \infty)$.

6.1.3. Path-connectedness.

DEFINITION. *Let X be a topological space. We say that X is **path-connected** if for all $x, y \in X$, there exists a continuous function $f : [0, 1] \to X$ such that $f(0) = x$ and $f(1) = y$.*
*This continuous function is usually called a **path** from x to y.*

REMARK. It must be remembered that $[0, 1]$ in the previous definition is always equipped with its usual topology.

Here is a quite practical lemma

LEMMA. *Let X be a topological space in which f is a path joining x and y, and g is a path joining y and z. Then the function*

$$h(t) = \left\{ \begin{array}{ll} f(2t), & 0 \le t \le \frac{1}{2}, \\ g(2t - 1), & \frac{1}{2} \le t \le 1, \end{array} \right.$$

defines a path joining x and z.

EXAMPLES.

(1) $\mathbb{R}$, *and in general,* $\mathbb{R}^n$ *($n \ge 1$) are path-connected in the standard topology.*
(2) $\mathbb{R}^*$ *is not path-connected.*

The notion of convexity gives more explicit examples of path-connected sets:

DEFINITION. *Let X be a real vector space and let $A \subset X$. We say that A is **convex** if:*

$$\forall x, y \in A, \forall t \in [0, 1] : \ (1 - t)x + ty \in A.$$

EXAMPLES.

(1) *A (real) linear subspace is convex.*
(2) $\mathbb{R}$ *is convex.*
(3) *Intervals are convex.*
(4) *The closed unit ball is convex.*

Thanks to the next result, the previous four examples are all path-connected:

THEOREM. *Any convex set is path-connected.*

The next result shows the relationship between connectedness and path-connectedness

THEOREM. *Every path-connected space is connected.*

REMARK. The converse of the previous theorem is not always true. See Exercise 6.5.7.

Nonetheless, there are sufficient conditions making a connected space path-connected (in usual $\mathbb{R}^n$), one of them is openness. We have:

THEOREM. *Every connected open subset of $\mathbb{R}^n$ is path-connected.*

6.2. True or False: Questions

QUESTIONS. Comment on the following questions/statements and indicate those which are false and those which are true when this applies. Justify your answers.

(1) Let T and T' be two topologies on the same set X. Assume that $T \subset T'$. Then

$$(X, T) \text{ is connected } \iff (X, T') \text{ is connected.}$$

(2) A closed subspace of a connected space is connected.

(3) The boundary of a connected set is itself connected.

(4) If the closure of some set is connected, then this set has to be connected. What about the converse?

(5) The union of connected sets is always connected.

(6) The intersection of connected sets is always connected.

(7) If a set A is connected, then so is its interior $\overset{\circ}{A}$.

(8) Let $f \colon [-1, 1] \to \mathbb{R}$ be a continuous function such that $f(-1)f(1) < 0$. Then there exists some $\alpha \in [-1, 1]$ such that $f(\alpha) = 0$.

(9) Every polynomial on $\mathbb{R}$ having an odd degree has at least one real root.

(10) If a set $\overline{A}$ is path-connected, then so is A.

(11) In the usual topology, $\mathbb{R}$ is homeomorphic to $\mathbb{R}^2$.

(12) The preimage of a connected set by a continuous function is connected.

(13) If X and Y are homeomorphic, then there is a one-to-one correspondence between their components.

(14) The quotient of a connected (path-connected respectively) topological space is connected (path-connected respectively).

(15) Let X be a topological space. Let C_x be the component of $x \in X$. Then

$$X \text{ is connected } \iff C_x = X.$$

(16) $\mathbb{R}^*$ is not connected as it has two components $(0, \infty)$ and $(-\infty, 0)$. But it is known from the lecture that the components are closed whereas here they are not closed in $\mathbb{R}$. Is there anything wrong with this reasoning?

6.3. Exercises With Solutions

Exercise 6.3.1. Is A connected in the topological space T in the following cases?

(1) $A = \{1, 2, 3, 4\}$, $T = \{\emptyset, \{1\}, \{2, 3\}, \{1, 2, 3\}, A\}\}$;
(2) A is a subset of some set endowed with the discrete topology (denoted by T);
(3) $A = \mathbb{R}$ and T the associated co-finite topology;
(4) $A = B((0, 1), 1) \cup B((0, -1), 1)$,
(5) $A = B_c((0, 1), 1) \cup B_c((0, -1), 1)$;
(6) $A = B((0, 1), 1) \cup B_c((0, -1), 1)$;
(the last three sets in $\mathbb{R}^2$ with respect to the euclidian metric).

Exercise 6.3.2. Let $X = \mathcal{M}_n(\mathbb{R})$, the (vector) space of square matrices of order n with real entries. Let A be the subset of X of invertible matrices. Is A connected?

Exercise 6.3.3. Show that $\mathbb{Q}$ is not connected using different methods.

Exercise 6.3.4. In the induced usual topology, are the following sets connected

(1) $\mathbb{R} \setminus \mathbb{Q}$;
(2) $\{\frac{1}{n} : n \geq 1\}$;
(3) $[0, 1) \cup (1, 2)$;
(4) $\mathbb{N}$?

Exercise 6.3.5. Show that $[0, 2)$ is connected with respect to the topology of Exercise 3.3.27.

Exercise 6.3.6. What are the components of

(1) $A = \mathbb{C} \setminus \mathbb{R}$;
(2) $A = \{(x, y) \in \mathbb{R}^2 : x \neq y\}$;
(3) $A = B((0, 1), 1) \cup B((0, -1), 1)$, $B = B_c((0, 1), 1) \cup B_c((0, -1), 1)$ in the usual $\mathbb{R}^2$?

Exercise 6.3.7. Let A be a connected set in a topological space X.

(1) If $A \subset B \subset \overline{A}$, then show that B is connected.
(2) Deduce that if A is connected, then so is $\overline{A}$.

Exercise 6.3.8. Let $\mathbb{R}$ be endowed with the K-topology.

(1) Show that $(-\infty, 0)$ and $(0, \infty)$ inherit their usual topology as subspaces of $\mathbb{R}_K$.

(2) Deduce that $\mathbb{R}$ is connected in $\mathbb{R}_K$.

Exercise 6.3.9. Show that there is no continuous function $f : \mathbb{R} \to \mathbb{R}$ such that

$$f(\mathbb{Q}) \subset \mathbb{R} \setminus \mathbb{Q} \text{ and } f(\mathbb{R} \setminus \mathbb{Q}) \subset \mathbb{Q}.$$

Exercise 6.3.10. Let $\mathbb{N}$ be endowed with the co-finite topology. Show that $\mathbb{N}$ is not path-connected.

REMARK. Do not say that since $\mathbb{N}$ is not an interval, it is not connected and hence it cannot be path-connected as every path-connected set is connected. This is a *wrong* reasoning for the simple fact that the connected sets are intervals in $\mathbb{R}$ **endowed with its usual topology**!

Exercise 6.3.11. Using *only* the definition of a path-connected set, show that $\mathbb{R}^*$ is not path-connected.

Exercise 6.3.12.

(1) Show that every convex part is path-connected.

(2) Deduce that $\mathbb{R}^n$ and the closed and open balls on $\mathbb{R}^n$ are all connected $(n \geq 1)$.

Exercise 6.3.13. Let $n \geq 1$.

(1) Is $\mathbb{R}^n \setminus \{0\}$ path-connected?

(2) Let

$$S^{n-1} = \{x \in \mathbb{R}^n : x^2 = 1\}.$$

Need S^{n-1} be path-connected?

Exercise 6.3.14. Let $\mathbb{T} = S(O_{\mathbb{R}^2}, 1)$ be the unit sphere in $\mathbb{R}^2$. Let A be the annulus on $\mathbb{R}^2$, i.e. $A = \{(x, y) \in \mathbb{R}^2 : a^2 \leq x^2 + y^2 \leq b^2\}$ where $0 < a < b$.

(1) Verify that the following functions are continuous $f : [a, b] \times \mathbb{T} \to A$ and $g : A \to [a, b] \times \mathbb{T}$ defined as

$$f(z, x, y) = (zx, zy) \text{ and } g(x, y) = \left(\sqrt{x^2 + y^2}, \frac{(x, y)}{\sqrt{x^2 + y^2}} \right).$$

Compute $f \circ g$ and $g \circ f$. What can you deduce from this question?

(2) Deduce from the previous equation that the annulus is a path-connected part of $\mathbb{R}^2$.

Exercise 6.3.15. Let X and Y be two topological spaces. Let $f : X \to Y$ be a homeomorphism. Let $a \in X$.

(1) Show that $\overline{f} : X \setminus \{a\} \to Y \setminus \{f(a)\}$ is a homeomorphism.
(2) Deduce that there cannot be a homeomorphism between $\mathbb{R}$ and $\mathbb{R}^2$.

Exercise 6.3.16. Consider the letters X and S as subsets of $\mathbb{R}^2$. Can we say that X and S are homeomorphic? What about E and W?

6.4. Tests

Test 46. Show that a topological space with one point is always connected. What about a set with two points?

Test 47. Is $\mathbb{R}$ connected with respect to the lower limit topology of $\mathbb{R}$?

Test 48. Show that $\mathbb{R}_\ell$ is totally disconnected.

Test 49. Is $\mathbb{R}$ connected with respect to the co-countable topology? What about $\mathbb{Q}$?

Test 50. Let X be a topological space. Suppose that X contains a connected dense subspace. Show that X is connected.

Test 51. What are the components of $\mathbb{R}$ with the respect to the co-finite topology?

Test 52. In the usual topology, can $[-1, 1]$ be homeomorphic to the unit circle?

6.5. More Exercises

Exercise 6.5.1. Prove that connectedness is a topological property.

Exercise 6.5.2. Set

$$A = \{\mathbb{R}\} \cup \{U \subset \mathbb{R} : U^c \text{ connected with respect to usual } \mathbb{R}\}.$$

Is T a topology on $\mathbb{R}$?

Exercise 6.5.3. Show that there does not exist any continuous function on the circle $\mathbb{T}$ into $\mathbb{R}$. Deduce from this that $\mathbb{T}$ is not homeomorphic to any subspace of $\mathbb{R}$ and hence $\mathbb{R}^2$ and $\mathbb{R}$ are not homeomorphic.

Exercise 6.5.4. Let X be a connected space. Assume that $\mathrm{card} X \geq 2$. Show that if for each $x \in X$, $\{x\}$ is closed, then $\mathrm{card} X = \infty$.

Exercise 6.5.5. Using Exercise 5.5.17, give an alternative way of proving that the unit circle is connected.

Exercise 6.5.6. Is $\mathbb{R}$ connected with respect to the topology of Exercise 5.3.13?

Exercise 6.5.7 (Important). In the euclidean metric, let

$$A = \left\{(x,y) \in \mathbb{R}^2 : 0 < x \leq 1, \ y = \sin\left(\frac{\pi}{x}\right)\right\}.$$

(1) Show that A is connected.
(2) Prove that $\overline{A}$ is not path-connected.

Exercise 6.5.8. In the usual topology of $\mathbb{R}^2$, say whether M and N are homeomorphic? What about B and V? or Y and T?

Exercise 6.5.9. Let

$$A = (\mathbb{R} \times \mathbb{Q}) \cup (\mathbb{Q} \times \mathbb{R}).$$

Show that A is path-connected.

Exercise 6.5.10. How many components does a puzzle of 2000 pieces have?

Exercise 6.5.11. Prove that the Cantor set is totally disconnected.

Exercise 6.5.12. Let X be a topological space. We say that X is **locally connected** at a point $x \in X$ if for each neighborhood of U of x, there exists a *connected* neighborhood V of x such that $V \subset U$.

(1) In the usual topology, are $\mathbb{R}$, $\mathbb{R}^*$ and $\mathbb{Q}$ locally connected?
(2) Show that the every open set in a locally connected set remains locally connected.
(3) Prove that the components of a locally connected space are open.

CHAPTER 7

Complete Metric Spaces

7.1. What You Need to Know

7.1.1. Completeness.

DEFINITION. *Let (X, d) be a metric space and let (x_n) be a sequence in X. We say that (x_n) is **Cauchy** if:*

$$\forall \varepsilon > 0, \exists N \in \mathbb{N}, \forall n, m \in \mathbb{N} : (n, m \geq N \implies d(x_n, x_m) < \varepsilon).$$

REMARK. Let (X, d) be a metric space and let $A \subset X$ be equipped with the induced metric, denoted by d_A. Let (x_n) be a sequence in A. Then (x_n) is Cauchy in (X, d) iff (x_n) is Cauchy in (A, d_A).

A convergent sequence is a metric space is Cauchy, but not vice versa. In fact, we have

DEFINITION. *A metric space (X, d) is said to be **complete** if every Cauchy sequence converges in (X, d), i.e. its limit belongs to X.*

EXAMPLES.
  (1) $(\mathbb{Q}, |\cdot|)$ *is not complete (see Exercise 7.3.1).*
  (2) $(\mathbb{R}, |\cdot|)$ *is complete (see Exercise 7.3.4).*
  (3) $(\mathbb{C}, |\cdot|)$ *is complete (see Exercise 7.3.5).*
  (4) $([0, 1), |\cdot|)$ *is not complete.*
  (5) *Any discrete metric space is complete (see Exercise 7.3.2).*

REMARK. If (X, d) and (X, d') are two metric spaces such that d and d' are equivalent, then it may be shown that (X, d) is complete iff (X, d') is complete.

If d and d' are (only) topologically equivalent, then the completeness of (X, d) need not imply that of (X, d'). For instance in $\mathbb{R}$, consider the usual metric d and the metric d' of Exercise 2.3.26 (which are not topologically equivalent). Then $(\mathbb{R}, d)$ is complete while $(\mathbb{R}, d')$ is not (see Test 54).

The next result illustrates the relationship between the notions of closedness and completeness.

THEOREM. *Let (X, d) be a metric space and let $A \subset X$. Then:*

(1) *If A is complete, then it is closed in X.*
(2) *If A is closed and X is complete, then A is complete.*

REMARK. Combining the two results in the preceding theorem, we may state with ease that: *In a complete metric space, a set is complete if and only if it is closed.*

The coming theorem gives a relationship between the concepts of compactness and completeness.

THEOREM. *Let X be a metric space.*

(1) *If X is compact, then it is complete.*
(2) *If X is totally bounded and complete, then it is compact.*

REMARK. We may then easily show that if X is a complete metric space, then $A \subset X$ is compact iff A is closed and totally bounded.

The next theorem is very important.

THEOREM. *Let (X, d) and (Y, d') be two metric spaces. Assume that $A \subset X$ is dense and that (Y, d') is complete. Let $f : (A, d) \to (Y, d')$ be uniformly continuous on A. Then there exists a unique function $g : (X, d) \to (Y, d')$ which is also uniformly continuous and such that $g(x) = f(x)$ for each $x \in A$.*

REMARK. The completeness of (Y, d) and the uniform continuity of f may not merely be dropped. See the section "True or False" for counterexamples and further discussion.

The next two theorems are also fundamental.

THEOREM (Cantor). *Let (X, d) be a metric space. Then (X, d) is complete iff for each sequence (A_n) of non-empty and closed subsets of X verifying:*

(1) $A_{n+1} \subset A_n$ *for all n;*
(2) $d(A_n) \to 0$ *as n goes to ∞;*

we have $\displaystyle\bigcap_{n=1}^{\infty} A_n \neq \varnothing.$

REMARK. In some textbooks, they refer to Cantor's theorem as the "principle of nested closed sets".

THEOREM (Baire). *Let (X, d) be a complete metric space. The countable union of closed sets with empty interior is also of empty interior.*

REMARK. We may also say that a topological space is a **Baire space** if the countable union of closed sets with empty interior is also of empty interior.

REMARK. Baire's theorem has important applications in functional analysis.

An equivalent version of Baire's theorem is the following:

THEOREM. *Let (X, d) be a complete metric space. The countable intersection of open dense sets is dense.*

7.1.2. Fixed Point theorem.

DEFINITION. *Let (X, d) be a metric space. A function $f : (X, d) \to (X, d)$ is called a **contracting mapping** (or a **contraction**) if:*

$$\exists k \in [0, 1) : \ d(f(x), f(y)) \leq kd(x, y).$$

PROPOSITION. *A contraction is uniformly continuous, hence continuous.*

DEFINITION. *Let X be a set and let $f : X \to X$ be a function. We say that x is a **fixed point** of f if $f(x) = x$.*

The next is a very interesting result in analysis:

THEOREM (Banach). *Let (X, d) be a complete metric space. Let $f : (X, d) \to (X, d)$ be a contracting mapping. Then f has a unique fixed point.*

REMARK. We have already met a somehow similar result for a "strict contraction". See Exercise 5.3.27

7.2. True or False: Questions

QUESTIONS. Comment on the following questions/statements and indicate those which are false and those which are true when this applies. Justify your answers.

(1) Let (X, d) be a metric space. It is known that (x_n) is a Cauchy sequence if

$$\forall \varepsilon > 0, \exists N \in \mathbb{N}, \forall n, m \in \mathbb{N} \ (n, m \geq N \Rightarrow d(x_n, x_m) < \varepsilon).$$

We can take $\varepsilon \geq 0$ and $d(x_n, x_m) \leq \varepsilon$ in the previous definition.

(2) How do we show that a given metric space is complete?

(3) How do we show that a given metric space is *not* complete?

(4) If a given space is complete, then what is the best way to exploit this hypothesis?

(5) Why is it important to consider complete spaces?

(6) $(0, \infty)$ is not complete.

(7) In a complete metric space, the intersection of two dense sets is never empty.

(8) The union of two non-complete sets can be complete.

(9) Let (X, d) be a metric space and let (x_n) be a sequence in (X, d). Then

$$(x_n) \text{ is Cauchy iff it is bounded.}$$

(10) Let (X, d) and (X, d') be two metric spaces. Assume that d and d' are equivalent. Let (x_n) be a sequence in X. Then is true that (x_n) is Cauchy with respect to d if and only if (x_n) is Cauchy with respect to d'?

(11) Let (X, d) and (X, d') be two metric spaces. Assume that d and d' are not equivalent. Let (x_n) be a sequence in X. Can (x_n) be Cauchy with respect to d but not so with respect to d' and vice versa?

(12) Let $X = C([0, 1], \mathbb{R})$ endowed with the *metric*

$$d(f, g) = \int_0^1 |f(x) - g(x)| dx.$$

Let

$$f_n(x) = \begin{cases} 0, & 0 \le x \le \frac{1}{2} - \frac{1}{n} \\ nx + (1 - \frac{1}{2}n), & \frac{1}{2} - \frac{1}{n} \le x \le \frac{1}{2} \\ 1, & \frac{1}{2} \le x \le 1. \end{cases}$$

This sequence has as its pointwise limit the function

$$f(x) = \begin{cases} 0, & 0 \le x < \frac{1}{2}, \\ 1, & \frac{1}{2} \le x \le 1. \end{cases}$$

Criticize the following reasoning: The sequence (f_n) is a Cauchy sequence of continuous functions. Since f is not continuous, (X, d) is not complete.

(13) Keeping everything as in the previous question, say whether the following reasoning is correct: The sequence (f_n) is a Cauchy sequence of continuous functions. Since f is not continuous and $d(f_n, f) \to 0$, (X, d) is not complete.

(14) Cauchyness is a topological property.

(15) Cauchyness is conserved by uniform continuity.

(16) Completeness is preserved by uniform continuity.

(17) Completeness is a topological property.

(18) Let X and Y be two metric spaces and let $A \subset X$ be dense in X. Let $f : A \to Y$ be a continuous function. Then f has a continuous extension $\tilde{f} : X \to Y$ such that $\tilde{f}_{|A} = f$.

(19) Let (X, d) be a metric space. State some properties that illustrate the analogy which exists between compact and complete spaces.

7.3. Exercises With Solutions

Exercise 7.3.1. Among the following sets indicate those which are complete (the metric is the usual one)

(1) $\mathbb{Q}$;
(2) $\mathbb{R} \setminus \mathbb{Q}$;
(3) $\mathbb{Q} \cap [3, 4]$;
(4) $(0, +\infty)$;
(5) $\{n : n \geq 1\}$;
(6) $\{(-1)^n : n \geq 1\}$;
(7) $\{\frac{1}{n} : n \in \mathbb{N}\} \cup \{0\}$;
(8) $\{(x, y) \in \mathbb{R}^2 : x > 1, y \geq \frac{1}{x-1}\}$?

Exercise 7.3.2. Show that a discrete metric space is always complete.

REMARK. Since every closed subset of a complete metric space is itself complete, since by this exercise discrete metric spaces are complete and since every subset of a discrete metric space is closed, we can state with ease that all subsets are complete in a discrete metric space (a propriety that characterizes discrete metric spaces).

Exercise 7.3.3 (Important). Let (X, d) be a metric space. Let (x_n) be a Cauchy sequence in X. Show that if (x_n) has a subsequence converging to $x \in X$, then (x_n) converges to $x \in X$ too.

Exercise 7.3.4. Show that $(\mathbb{R}, | \cdot |_{\mathbb{R}})$ is complete.

Exercise 7.3.5. Show that $(\mathbb{C}, | \cdot |_{\mathbb{C}})$ is complete.

Exercise 7.3.6.

(1) Show that every compact metric space is complete.
(2) Is the converse always true?
(3) Deduce from the first question another proof of the completeness of usual $\mathbb{R}$.
(4) Can we say that every complete space is locally compact and vice versa?

Exercise 7.3.7. Define the following *metric* on $\mathbb{R}$

$$(x,y) \mapsto d(x,y) = \left| \frac{x}{1+|x|} - \frac{y}{1+|y|} \right|.$$

Show that $(\mathbb{R}, d)$ is not complete (hint: you may use the sequence defined by $x_n = n$, $n \in \mathbb{N}$).

Exercise 7.3.8. Show that $(\mathbb{N}, d)$, the metric space defined in Exercise 2.5.11, is complete.

Exercise 7.3.9. Let

$$\ell^2 = \left\{ x = (x_n)_n, \, x_n : \mathbb{N} \to \mathbb{C} : \sum_{n=1}^{\infty} |x_n|^2 < +\infty \right\}.$$

Define a function d on $\ell^2 \times \ell^2$ by

$$d(x,y) = \sqrt{\sum_{n=1}^{\infty} |x_n - y_n|^2}.$$

(1) Give some elements in ℓ^2 and others not in it.
(2) Show that (ℓ^2, d) is a complete metric space

Exercise 7.3.10. We endow $\mathbb{R}$ with following *metric*

$$d(x,y) = |e^x - e^y|.$$

Show that $\mathbb{R}$ is not complete with respect to d (hint: you may use the sequence defined by $x_n = -n$, $n \in \mathbb{N}$).

Exercise 7.3.11 (Classic and Important). In (ℓ^2, d) as defined in Exercise 7.3.9, show that

$$A = \{(x_n)_n \in \ell^2 : \exists N \in \mathbb{N}^*, \, x_n = 0, \forall n \geq N\}$$

is not complete.

Exercise 7.3.12. Let $X = C([0,1], \mathbb{R})$. We equip it with the *metric* defined, for all $f, g \in X$, by

$$d(f,g) = \int_0^1 |f(x) - g(x)| dx.$$

Consider the sequence of *continuous* functions

$$f_n(x) = \begin{cases} 0, & 0 \leq x \leq \frac{1}{2} - \frac{1}{n} \\ nx + (1 - \frac{1}{2}n), & \frac{1}{2} - \frac{1}{n} \leq x \leq \frac{1}{2} \\ 1, & \frac{1}{2} \leq x \leq 1. \end{cases}$$

(1) Show that (f_n) is a Cauchy sequence.
(2) Is (X, d) complete?

Exercise 7.3.13. [classic and important] Let $X = C([0, 1])$. Show that (X, d_∞) is complete where d_∞ is the supremum metric, i.e. the metric defined for all $f, g \in X$ by

$$d_\infty(f, g) = \sup_{x \in [0,1]} |f(x) - g(x)|$$

(hint: you need the fact that the uniform limit of a sequence of continuous functions is continuous. This should be known to the reader, if not, then more details are to be found in the next chapter).

Exercise 7.3.14. Let X be the space of bounded real-valued functions on $[0, 1]$. Endow X with the supremum metric which we denote by d. Show that (X, d) is a complete metric space.

Exercise 7.3.15. Let X be the set of all polynomials (of any degree) defined on $[0, 1]$ and let d be the metric defined by

$$d(f, g) = \sup_{x \in [0,1]} |f(x) - g(x)|, \ f, g \in X.$$

(1) Show that the function $x \mapsto e^x$ is not a polynomial.
(2) Using the sequence $P_n(x) = \left(1 + \frac{x}{n}\right)^n$, $n \geq 1$, show that (X, d) is not complete.
(3) Give another proof of the non-completeness of (X, d) using the Weierstrass theorem.

Exercise 7.3.16. Let $A = (0, \infty)$. Let d be the usual metric and for any $x, y \in A$, define the function (cf. Exercise 2.5.13)

$$d'(x, y) = |\ln x - \ln y|.$$

(1) Show that d' is a metric on A.
(2) Why is (A, d) not complete?
(3) Show that (A, d') is complete.
(4) Show that (A, d) is homeomorphic to (A, d').
(5) What can you deduce from the previous question? (cf. Exercise 2.5.13).

Exercise 7.3.17.

(1) Show that every differentiable function in $[a, b]$ into $[a, b]$, whose derivative is bounded by some $M < 1$, is a contraction.
(2) Show, using the fixed point theorem, that the equation

$$x^4 + 16x^3 - 32x - 8 = 0$$

has exactly one root in $[0, \frac{1}{2}]$.

Exercise 7.3.18.

(1) Let X be a complete metric space. Let f be a mapping defined on X into X. We assume that for some integer $n \geq 1$, f^n is a contraction. Show then that f has a unique fixed point.

(2)   • Show that $x \mapsto \cos^2 x$ is a contraction while $x \mapsto \cos x$ is not one.

     • Give an approximate solution of the equation $\cos x = x$.

Exercise 7.3.19. Let $X = \{x \in \mathbb{Q} : x \geq 1\}$ in the usual metric. Define a function on X into X by

$$f(x) = \frac{x}{2} + \frac{1}{x}.$$

(1) Show that

$$\forall x, y \in X : |f(x) - f(y)| \leq \frac{1}{2}|x - y|.$$

(2) Show that f does not have a fixed point, that is,

$$\nexists x \in X : f(x) = x.$$

(3) Why does not the result of the previous question contradict the fixed point theorem?

Exercise 7.3.20. (cf. Exercise 5.3.27) Define a function f in $X = [1, \infty)$ (with the usual metric) into the same interval by

$$f(x) = x + \frac{1}{x}.$$

(1) Show that

$$\forall x, y \geq 1 \ (x \neq y) : |f(x) - f(y)| < |x - y|.$$

(2) Show that f does not have a fixed point, that is,

$$\forall x \in X : f(x) \neq x.$$

(3) Why does the result of the previous question not contradict the fixed point theorem? Say why it does not contradict the result of Exercise 5.3.27 either.

Exercise 7.3.21. Let (X, d) and (X', d') be two isometric metric spaces. Show that (X, d) is complete if and only if (X', d') is complete.

Exercise 7.3.22. Go back to Cantor's theorem and give examples showing that the hypotheses on the right hand side of the equivalence may not merely be dropped.

Exercise 7.3.23. Let $a, b \in \mathbb{R}$ with $a \leq b$. Show that $[a, b]$ is compact using total boundedness.

Exercise 7.3.24.

(1) Show that $\mathbb{R}$ is not countable (hint: use Baire's theorem).
(2) Deduce that $\mathbb{R} \setminus \mathbb{Q}$ is also uncountable.

Exercise 7.3.25. Let C be a non-empty closed and countable subset of usual $\mathbb{R}$. Show that C has at least one isolated point.

Exercise 7.3.26 (Important). Assume that (X, d) and (Y, d') are two metric spaces. Assume also that (X, d) is complete. Let A be closed and let $f : A \to Y$ be a continuous function satisfying:

$$d'(f(x), f(x')) \geq d(x, x'), \ \forall x, x' \in A.$$

Prove that $f(A)$ is closed (Y, d').

7.4. Tests

Test 53. Let X be a metric space and let $A \subsetneq X$ be dense in X. Must A be complete?

Test 54. Show that $\mathbb{R}$ endowed with the "arctan metric" already defined in Exercise 2.3.26 is not complete.

Test 55. Show that $[0, 1)$ is not countable.

Test 56. Show that the Cantor set is uncountable.

7.5. More Exercises

Exercise 7.5.1. Set

$$x_n = 1 + \frac{1}{2} + \cdots + \frac{1}{n} = \sum_{k=1}^{n} \frac{1}{k}.$$

Show that (x_n) is not Cauchy in $\mathbb{R}$. Does the series $\sum_{n \geq 1} \frac{1}{n}$ diverge?

Exercise 7.5.2. Is the Cantor set complete?

Exercise 7.5.3. Let (X, d) be a *ultrametric* space (see Exercise 2.3.20). Let (x_n) be a sequence (x_n) in X. Show that

$$(x_n) \text{ is Cauchy} \iff \lim_{n \to \infty} d(x_n, x_{n+1}) = 0.$$

Exercise 7.5.4. For all $(x, y) \in \mathbb{R}^2$, define the *metric* d by

$$d(x, y) = |x^3 - y^3|.$$

(1) Show that $\mathbb{R}$ is complete with respect to d.

(2) Do the previous question with the metric d defined by

$$d(x, y) = \left| \frac{1}{x} - \frac{1}{y} \right|$$

and $\mathbb{R}^*$ in lieu of $\mathbb{R}$.

Exercise 7.5.5. Let $X = C([-1, 1], \mathbb{R})$. We endow X with the *metric* d defined for all $f, g \in X$ by

$$d(f, g) = \left(\int_{-1}^{1} |f(x) - g(x)|^2 dx. \right)^{\frac{1}{2}}$$

and let

$$f_n(x) = \begin{cases} -1, & -1 \le x \le -\frac{1}{n}, \\ nx, & -\frac{1}{n} \le x \le \frac{1}{n}, \\ 1, & \frac{1}{n} \le x \le 1. \end{cases}$$

(1) Show that (f_n) is a Cauchy sequence in (X, d).
(2) Is (X, d) complete?

Exercise 7.5.6. We endow $X = [0, 1)$ with the following *metric*

$$d(x, y) = \left| \frac{1}{1 - x} - \frac{1}{1 - y} \right|.$$

Is (X, d) complete?

Exercise 7.5.7. In (ℓ^∞, d) where

$$\ell^\infty = \{x = (x_n), \ x_n : \mathbb{N} \to \mathbb{C}, \ \sup_n |x_n| < \infty\}$$

and

$$d(x, y) = \sup_{n \in \mathbb{N}} |x_n - y_n|, \ \forall x, y \in A,$$

show that

$$A = \{(x_n)_n \in \ell^\infty : \exists N \in \mathbb{N}^*, \ x_n = 0, \forall n \ge N\}$$

is not complete.

Exercise 7.5.8. Let $f : (\mathbb{N}, d) \to (\mathbb{N}, d)$ be a function defined for all $x \in \mathbb{N}$ by $f(x) = x + 1$, where d is the metric defined in Exercise 2.5.11.

(1) Verify that $(\mathbb{N}, d)$ is complete.
(2) Without calculation, can f be a contraction? Why?

Exercise 7.5.9. Show that the image of a Cauchy sequence by a uniform continuous function remains Cauchy.

Exercise 7.5.10. Let $f : \mathbb{R}^2 \to \mathbb{R}^2$ be defined by

$$f(x, y) = (\frac{1}{2} \cos y, \frac{1}{2} \sin x + 1)$$

where $\mathbb{R}^2$ is equipped with the usual metric. Show that f has a unique fixed point.

Exercise 7.5.11. Define $f : \mathbb{R} \to \mathbb{R}$ (usual $\mathbb{R}$!) by $f(x) = e^{-x}$. Show that f is not a contraction yet it has a unique fixed point (hint: consider f^2).

Exercise 7.5.12. Let (X_1, d_1) and (X_2, d_2) be two complete metric spaces. Endow the product space $X = X_1 \times X_2$ with the *metric d'* defined by

$$d'(x, y) = \max_{1 \le i \le 2} d_i(x, y)$$

for all $x, y \in X$. Show that (X, d) is complete.

Exercise 7.5.13. Let (f_n) the sequence defined in Exercise 4.3.27. Is $\{f_n : n \in \mathbb{N}\}$ totally bounded?

Exercise 7.5.14. Let (X, d) be a metric space. Recall the following important result (see, for instance, [**16**]):

THEOREM. *(Hausdorff) There exist a complete metric space $(\tilde{X}, \tilde{d})$ and an isometry $f : X \to \tilde{X}$ such that $f(X)$ is dense in $\tilde{X}$.*

The space $(\tilde{X}, \tilde{d})$ is usually called a **completion** of (X, d).

(1) Let $A \subset X$. If X is complete, then what is the completion of A?
(2) What is the completion of $(-1, 1)$ and that of $\mathbb{Q}$ in usual $\mathbb{R}$?
(3) Explain why the completion of the space of polynomials on $[0, 1]$ with respect to the supremum metric is the space of continuous functions on $[0, 1]$ with respect to the supremum metric.

CHAPTER 8

Function Spaces

8.1. What You Need to Know

8.1.1. Types of Convergence.

DEFINITION. *Let (X, d) and (Y, d') be two metric spaces. Let (f_n) be a sequence of functions defined on X into Y. We say that (f_n) converges **pointwise** (or **simply**) to $f : X \to Y$ if:*

$$\lim_{n \to \infty} f_n(x) = f(x) \quad (in \ \mathbb{R}).$$

DEFINITION. *Let X be a topological space and let (Y, d') be a metric space. Let (f_n) be a sequence of functions defined on X into Y. We say that (f_n) converges **uniformly** to $f : X \to Y$ if:*

$$\forall \varepsilon > 0, \exists N \in \mathbb{N}, \forall x \in X : (n \geq N \implies d'(f_n(x), f(x)) < \varepsilon).$$

REMARK. If f_n is real-valued, defined on $A \subset \mathbb{R}$ and converges pointwise to f, then saying that (f_n) converges to f uniformly amounts to saying that

$$\lim_{n \to \infty} \sup_{x \in A} |f_n(x) - f(x)| = 0.$$

EXAMPLES.

(1) *Let $f_n(x) = x^n$ where $x \in [0, 1]$. Then (f_n) converges pointwise to*

$$f(x) = \begin{cases} 0, & 0 \leq x < 1, \\ 1, & x = 1. \end{cases}$$

(2) *The same sequence does not converge uniformly to f for:*

$$\lim_{n \to \infty} \sup_{0 \leq x \leq 1} |f_n(x) - f(x)| \geq \lim_{n \to \infty} \sup_{0 \leq x < 1} |x^n - 0| = 1$$

and hence $\lim_{n \to \infty} \sup_{0 \leq x \leq 1} |f_n(x) - f(x)| \neq 0$.

Here are some applications of uniform convergence. We content ourselves to results of sequences of functions defined on a subset of $\mathbb{R}$.

THEOREM. *Let (f_n) be a sequence of real-valued functions defined on $[a, b]$. Assume that (f_n) converges uniformly to a function f. Then:*

(1) (f_n) *converges pointwise to* f.
(2) *If all* (f_n) *are continuous, then* f *is continuous too.*
(3) *If all* (f_n) *are Riemann-integrable, then* f *is Riemann-integrable and*

$$\lim_{n\to\infty} \int_a^b f_n(x)dx = \int_a^b f(x)dx.$$

An akin result on differentiability needs a little more care.

THEOREM. *If* (f_n) *is a sequence of differentiable real-valued functions defined on* (a, b) *and which converges* **pointwise** *to* f, *and if the sequence of derivatives* (f'_n) *converges* **uniformly** *to a function* g *on* (a, b), *then* f *is differentiable and* $f' = g$.

Pointwise convergence may imply uniform convergence under extra hypotheses.

THEOREM (Dini). *Let* (f_n) *be an increasing sequence of real-valued continuous functions defined on (the compact!)* $[a, b]$. *Assume that* (f_n) *converges pointwise to a continuous function* f. *Then* (f_n) *converges uniformly to* f.

REMARK. By (f_n) increasing, we mean increasing with respect to n.

REMARK. The same conclusions of Dini's theorem hold:
(1) if "increasing" is replaced by "decreasing";
(2) or if $[a, b]$ is replaced by another compact set.

DEFINITION. *Let* X *be a topological space and let* (Y, d) *be a metric space. Let* (f_n) *be a sequence of functions from* X *into* (Y, d). *We say that* (f_n) **converges uniformly on compacta** *(or* **compactly***) to* $f : X \to (Y, d)$ *if:*

$\forall A$ *compact* $\subset X, \forall \varepsilon > 0, \exists N \in \mathbb{N} : (n \geq N \implies d(f_n(x), f(x)) < \varepsilon)$

for all $x \in A$.

REMARK. The "N" in the foregoing definition depends on both A and ε.

THEOREM. *Let* X *be a topological space and let* (Y, d) *be a metric space. Let* (f_n) *be a sequence of functions from* X *into* (Y, d).
(1) *If* (f_n) *converges uniformly to* f, *then* (f_n) *converges on compacta to* f.
(2) *Let* X *be compact. If* (f_n) *converges on compacta to* f, *then* (f_n) *converges uniformly to* f.

8.1.2. Weierstrass Approximation Theorem. The Weierstrass theorem has a wide range of applications in different areas of mathematics.

THEOREM (Weierstrass). *The subspace of real-valued polynomials defined on* $[a, b]$ *is dense in the space of real-valued continuous functions defined on* $[a, b]$*, with respect to the supremum metric.*

REMARK. In other words, the Weierstrass theorem tells us that any real-valued continuous function (on $[a, b]$) may be approximated uniformly by a sequence of real-valued polynomials (on $[a, b]$).

REMARK. There exist many generalizations of Weierstrass theorem. The most famous one is the Stone-Weierstrass theorem. For the reader's convenience we shall state here, but without further use in this book. But first we need to look at two notions:

DEFINITION. *Let* X *be a compact space equipped with the uniform metric. The subset* $A \subset C(X, \mathbb{R})$ *is called* **separating** *if*

$$\forall x, y \in X, x \neq y, \exists f \in A : f(x) \neq f(y).$$

DEFINITION. *A vector subspace* A *of* $C(X, \mathbb{R})$ *is a* **subalgebra** *if for all* $f, g \in A$*:* fg *(the pointwise product) remains in* A*.*

THEOREM (Stone-Weierstrass). *Let* X *be a compact space equipped with the uniform metric. Then every separating subalgebra of* $C(X, \mathbb{R})$ *which contains the constant functions is dense in* $C(X, \mathbb{R})$*.*

8.1.3. Arzelà-Ascoli Theorem. The epilogue of this chapter is devoted to the Arzelà-Ascoli theorem. But first we have the following definition:

DEFINITION. *A subset* $A \subset C[0, 1]$ *is said to be* **equicontinuous** *at a point* a *if:*

$$\forall \varepsilon > 0, \exists \alpha > 0, \forall f \in A, \forall x \in [0, 1] : (|x - a| < \alpha \implies |f(x) - f(a)| < \varepsilon).$$

A subset $A \subset C[0, 1]$ *is* **equicontinuous** *if it is equicontinuous at each point of* $[0, 1]$*.*

EXAMPLES.

(1) *A finite subset of* $C[0, 1]$ *is equicontinuous.*
(2) *The set of real-valued functions* f *defined on* $[0, 1]$ *and satisfying the Hölder condition:*

$$\exists k, \alpha > 0, \forall x, y \in [0, 1] : |f(x) - f(y)| \leq k|x - y|^{\alpha}$$

is equicontinuous.

In an infinite dimensional space, compact spaces are rare. For instance, the closed unit ball in $C[0,1]$ is not compact with respect to the supremum metric (see Exercise 5.3.20 or Exercise 5.5.11). The following theorem characterizes compact subsets of $C[0,1]$.

THEOREM (Arzelà-Ascoli). *Let $A \subset C[0,1]$ be closed and bounded. Then A is compact iff it is equicontinuous.*

REMARK. Of course, an alternative way of stating the previous result is to say: A bounded $A \subset C[0,1]$ is relatively compact iff it is equicontinuous.

REMARK. The Arzelà-Ascoli theorem is tremendously important in mathematics: It has applications in Ordinary Differential Equations, Distribution Theory and Holomorphic Functions, among others.

8.2. True or False: Questions

QUESTIONS. Comment on the following questions/statements and indicate those which are false and those which are true when this applies. Justify your answers.

(1) Let (f_n) be a monotonic sequence of continuous real-valued functions on $\mathbb{R}$. If (f_n) converges pointwise to f and f is continuous, then it converges uniformly to f.

(2) Let (f_n) be a sequence of continuous real-valued functions having f as its pointwise limit. If each function f_n is increasing, then so will be f.

(3) Let (f_n) be a sequence of bounded real-valued functions. Then its pointwise limit f is bounded. What about the uniform limit?

(4) Let (f_n) be a sequence of continuous real-valued functions that converges pointwise to f on $[0,1)$. If (f_n) converges uniformly on $[0,a]$ for each $0 < a < 1$, then the convergence is uniform on $[0,1)$.

(5) Let (f_n) be a sequence of continuous real-valued functions that converges pointwise to f on $(0,1]$. If f is continuous, then the convergence becomes uniform.

(6) Let (f_n) be a sequence of continuous real-valued functions defined on $[a,b]$ which converges pointwise to f. Then

$$\lim_{n\to\infty} \int_a^b f_n(x)dx = \int_a^b \lim_{n\to\infty} f_n(x)dx \quad \left(= \int_a^b f(x)dx \right).$$

(7) Let (f_n) be a sequence of continuous and differentiable real-valued functions that converges pointwise to f. Then

$$\lim_{n\to\infty} f_n'(x) = f'(x).$$

8.3. Exercises With Solutions

Exercise 8.3.1. Let (f_n) be the sequence of real-valued functions defined by $f_n(x) = \frac{\sin nx}{\sqrt{n}}$ for all $x \in [0,1]$.

(1) What is the pointwise limit of (f_n)?

(2) Does (f_n) converge uniformly to the pointwise limit? Why?

Exercise 8.3.2. Let (f_n) be the sequence of real-valued functions defined by $f_n(x) = \frac{e^{nx}}{\sqrt{n}}$ for all $x \in [-1,0]$.

(1) Show that (f_n) converges pointwise to some function f.

(2) Show that the convergence is also uniform.

(3) Study the uniform convergence of (f_n').

Exercise 8.3.3. Let (P_n) be a sequence of polynomials (on $[-1,1]$) defined by

$$\begin{cases} P_0 = 0, \\ P_{n+1}(x) = P_n(x) + \frac{1}{2}(x^2 - P_n^2(x)), \quad \forall n \geq 0. \end{cases}$$

Show that (P_n) converges uniformly to $f(x) = |x|$.

Exercise 8.3.4. Let (x^n) be a sequence of functions defined from $[0, \frac{1}{2})$ into $\mathbb{R}$; denote by (f_n). Investigate the convergence of (f_n):

(1) pointwise;

(2) uniformly;

(3) compactly.

The same questions for (g_n) defined from $\mathbb{R}$ into $\mathbb{R}$ by: $g_n(x) = \frac{n+1}{n}x$.

Exercise 8.3.5. Let (X, d) and (Y, d') be two metric spaces and let (f_n) be a sequence of equicontinuous functions from X into Y. Set

$$V = \{x \in X : (f_n(x)) \text{ is Cauchy in } Y\}.$$

Show that V is closed in Y.

Exercise 8.3.6. Let k be a continuous real-valued map on $[0,1] \times [0,1]$. Define a (linear) map K for all $f \in C([0,1])$ by

$$Kf(x) = \int_0^1 k(x,y)f(y)dy$$

(K is usual called an integral operator with kernel k). Let (f_n) be bounded.

(1) Show that (Kf_n) is equicontinuous.
(2) Using the Arzelà-Ascoli theorem, deduce that (Kf_n) is relatively compact in $C([0,1])$ (we then say that K is a **compact operator**).

Exercise 8.3.7. Let $f : [a,b] \to \mathbb{R}$ be a continuous function. The **moments** of f are the numbers

$$m(f) = \int_a^b x^n f(x)dx, \quad \text{for all } n = 1, 2, 3, \cdots$$

Show that if all the $m(f)$ vanish, then so will do f (hint: Use the Weierstrass approximation theorem).

Exercise 8.3.8. Endow $C([0,1], \mathbb{R})$ with the supremum distance. Show that $C([0,1], \mathbb{R})$ is separable (hint: show that the set of polynomials with rational coefficients is dense in $C([0,1], \mathbb{R})$).

8.4. Tests

Test 57. What is the pointwise limit (denoted by f) of the sequence of functions (f_n) defined by

$$f_n(x) = 1 + x + \cdots + x^n$$

on $(-1, 1)$? Does (f_n) converge uniformly to f?

Test 58. Let (f_n) be a sequence of real-valued functions defined on $[0, 1]$ by

$$f_n(x) = (x(1 - x))^n + x.$$

Find the pointwise limit and study the uniform convergence of (f_n).

Test 59. Let (P_n) be a polynomial sequence that converges uniformly to f. Is f continuous?

Test 60. Need a finite union of equicontinuous sets be equicontinuous?

8.5. More Exercises

Exercise 8.5.1. Define a sequence of functions (f_n) as follows

$$f_n(x) = f\left(x + \frac{1}{n^2}\right)$$

where $f : \mathbb{R} \to \mathbb{R}$ is a continuous function.
(1) Verify that (f_n) converges pointwise to f.
(2) Show that the convergence is also uniform.

(3) Deduce the value of $\displaystyle\lim_{n\to\infty} \int_0^1 f\left(x + \frac{1}{n^2}\right) dx.$

Exercise 8.5.2. The same questions as those of Exercise 8.3.4 for (f_n) defined from $\mathbb{R}$ into $\mathbb{R}$ by:

$$f_n(x) = \begin{cases} \frac{1}{n}\sqrt{n^2 - x^2}, & |x| < n, \\ 0, & \text{otherwise.} \end{cases}$$

Exercise 8.5.3. Let (X, d) and (Y, d') be two metric spaces and let (f_n) be a sequence of functions on X into Y assumed to be equicontinuous at $a \in X$. Let (x_n) be a sequence in X that converges to a. Show that if $(f_n(a))$ converges to b, then so does $(f_n(x_n))$.

Exercise 8.5.4. Let f be a uniformly continuous real-valued function on $\mathbb{R}$. Let $a \in \mathbb{R}$ and set $f_a(x) = f(x - a)$. Show that $\{f_a : a \in \mathbb{R}\}$ is equicontinuous on $\mathbb{R}$.

Exercise 8.5.5.

(1) Let X be a metric space and let (f_n) be a sequence in $C(X)$. Show that if (f_n) is an equicontinuous family at a point a, then for all sequence (x_n) converging to a, $(f_n(x) - f_n(x_n))$ converges to 0.

(2) What about the converse? (hint: consider $f_n(x) = \sin nx$ and $x_n = a + \frac{\pi}{2n}$).

Exercise 8.5.6. Let $f_n(x) = \sin(\sqrt{4(n\pi)^2 + x})$ be defined for all $x \geq 0$.

(1) What is the pointwise limit of (f_n)?
(2) Show that (f_n) is equicontinuous.
(3) Show that (f_n) is not relatively compact in $C([0, \infty), \mathbb{R})$ equipped with the supremum metric.
(4) Why does not the result of the previous question contradict the Arzelà-Ascoli theorem?

Part 2

Solutions

CHAPTER 1

General Notions: Sets, Functions et al

1.2. Solutions to Exercises

SOLUTION 1.2.1.

(1) We have

$$y \in f(\bigcup_{i \in I} A_i) \Longleftrightarrow \exists x \in \bigcup_{i \in I} A_i : y = f(x)$$
$$\Longleftrightarrow \exists i \in I,\ x \in A_i \text{ and } y = f(x)$$
$$\Longleftrightarrow \exists i \in I :\ y \in f(A_i)$$
$$\Longleftrightarrow y \in f(\bigcup_{i \in I} A_i).$$

(2) We have

$$\bigcap_{i \in I} A_i \subset A_j,\ \forall j \in I \Longrightarrow f(\bigcap_{i \in I} A_i) \subset f(A_j),\ \forall j \in I.$$

Thus

$$f(\bigcap_{i \in I} A_i) \subset \bigcap_{i \in I} f(A_i).$$

REMARK. Observe that Equality does not hold even for finite intersections. For instance, take the real-valued function f defined on $\mathbb{R}$ by $f(x) = x^2$. Then consider for example $A = (-\infty, 0]$ and $B = [0, \infty)$ and one can see easily that

$$f(A \cap B) = f(\{0\}) = \{0\} \neq [0, \infty) = f(A) \cap f(B),$$

i.e. the equality does not hold.

(3) We have

$$x \in f^{-1}(\bigcup_{i \in I} B_i) \Longleftrightarrow f(x) \in \bigcup_{i \in I} B_i \Longleftrightarrow \exists i \in I,\ f(x) \in B_i$$
$$\Longleftrightarrow \exists i \in I,\ x \in f^{-1}(B_i) \Longleftrightarrow x \in \bigcup_{i \in I} f^{-1}(B_i).$$

(4) The same proof as in the previous question.

(5) We have

$$x \in f^{-1}(B^c) \Longleftrightarrow f(x) \in B^c \Longleftrightarrow f(x) \notin B \Longleftrightarrow x \notin f^{-1}(B)$$
$$\Longleftrightarrow x \in [f^{-1}(B)]^c.$$

(6) We have

$$x \in (g \circ f)^{-1}(A) \Longleftrightarrow g(f(x)) \in A \Longleftrightarrow f(x) \in g^{-1}(A)$$
$$\Longleftrightarrow x \in f^{-1}(g^{-1}(A)).$$

(7) We have

$$y \in f(f^{-1}(A)) \Longrightarrow \exists x \in f^{-1}(A) : \; y = f(x) \Longrightarrow y = f(x) \in A.$$

(8) A similar method to that of the previous question works.

SOLUTION 1.2.2. Let $x \in A \cap f^{-1}(U)$. Then $x \in A$ and $x \in f^{-1}(U)$ or $f(x) \in U$. Since $x \in A$, we get $f_A(x) \in U$, i.e. $x \in f_A^{-1}(U)$.

Conversely, let $x \in f_A^{-1}(U)$ which, by definition, means that $x \in A$ and $f_A(x) \in U$. Hence $x \in A$ and $f(x) \in U$. Thus $x \in A$ and $x \in f^{-1}(U)$ or $x \in A \cap f^{-1}(U)$.

SOLUTION 1.2.3.

(1) We need only show that f is one-to-one. Let $(n, m), (n', m') \in \mathbb{N} \times \mathbb{N}$ such that $f(n, m) = f(n', m')$, i.e. $2^n 3^m = 2^{n'} 3^{m'}$. We need to show that $(n, m) = (n', m')$.

Suppose first that $n \neq n'$, then $2^n 3^m = 2^{n'} 3^{m'}$ implies $3^m = 2^{n'-n} 3^{m'}$ and if $n < n'$, then this is a contradiction as $2^{n'-n} 3^{m'}$ is even while 3^m is odd. If $n > n'$, then one considers $2^{n-n'} 3^m = 3^{m'}$ and the same contradiction will be encountered. Hence this forces us to have $n = n'$ and thus $3^m = 3^{m'}$. We can use a similar method to show that $m = m'$ or we can just take the logarithm to obtain $m = m'$. Thus f is one-to-one and hence $\mathbb{N} \times \mathbb{N}$ is countable.

(2) Since A and B are countable, there are *surjections* $f : \mathbb{N} \to A$ and $g : \mathbb{N} \to B$. Hence the map $h : \mathbb{N} \times \mathbb{N} \to A \times B$ defined, for all $(n, m) \in \mathbb{N} \times \mathbb{N}$ by $h(n, m) = (f(n); g(m))$ is a surjection. This leads to the countability of $A \times B$ since $\mathbb{N} \times \mathbb{N}$ is countable.

SOLUTION 1.2.4. Let $f : \bigcup_{n \geq 0} \mathbb{Q}^{n+1} \to \mathbb{Q}[X]$ be the map defined by

$$f(a_0, a_1, \cdots, a_n) = a_n x^n + \cdots + a_1 x + a_0.$$

By construction, f is onto. Since $\bigcup_{n \geq 0} \mathbb{Q}^{n+1}$ is countable (why?), so is $\mathbb{Q}[X]$.

SOLUTION 1.2.5. One way of proving that X is uncountable is to show that all countable subsets of X are proper, i.e. whenever a subset of X is countable, it will be strictly contained in X. Let $A \subset X$ be countable. Then there is some sequence of functions (f_n) (whose image consists of zeroes and ones) such that $A = \{f_n : n \in \mathbb{N}\}$. Define $f : \mathbb{N} \to \{0, 1\}$ by

$$f(n) = 1 - f_n(n), \ \forall n \in \mathbb{N}.$$

Now, if $f_n(n) = 1$, then $f(n) = 0$; and if $f_n(n) = 0$, then $f(n) = 1$. Hence $f \in A$. However, $f \notin A$ since if it were, then we would have

$$f(n) = 1 - f_n(n) = f_n(n),$$

which is absurd.

SOLUTION 1.2.6. The method of proofs is standard and it mainly uses the Archimedes theorem.

(1) Obviously $0 \in \left(\frac{-1}{n}, \frac{1}{n}\right)$ for all $n \geq 1$ and hence

$$\{0\} \subset \bigcap_{n \in \mathbb{N}} \left(\frac{-1}{n}, \frac{1}{n}\right).$$

Let $x \neq 0$. There exists N such that $|x| > \frac{1}{N} > 0$ (why?). Thus $x \notin \left(\frac{-1}{N}, \frac{1}{N}\right)$ and x is not in the intersection. This proves that $\bigcap_{n \in \mathbb{N}} \left(\frac{-1}{n}, \frac{1}{n}\right) \subset \{0\}$, establishing the equality.

(2) We always have $\bigcup_{n \in \mathbb{N}} [-n, n] \subset \mathbb{R}$. Now let $x \in \mathbb{R}$. Again by the Archimedes theorem we know that there exists $n \in \mathbb{N}$ such that $|x| \leq n$. Hence $x \in \bigcup_{n \in \mathbb{N}} [-n, n]$. Thus,

$$\bigcup_{n \in \mathbb{N}} [-n, n] = \mathbb{R}.$$

(3) To be shown as the preceding question.

SOLUTION 1.2.7. The reader may try to find the following results and must try to prove them as done in the foregoing exercise. The answers are

(1) $\bigcap_{n \in \mathbb{N}} A_n = \{1\}$ and $\bigcup_{n \in \mathbb{N}} A_n = \mathbb{N}$.

(2) $\bigcap_{n \in \mathbb{N}} A_n = (-1, 1)$ and $\bigcup_{n \in \mathbb{N}} A_n = \mathbb{R}$.

(3) $\bigcap_{n\in\mathbb{N}} A_n = [0,1]$ and $\bigcup_{n\in\mathbb{N}} A_n = (-1,2)$.

(4) $\bigcap_{n\in\mathbb{N}} A_n = \{0\}$ and $\bigcup_{n\in\mathbb{N}} A_n = [0,1)$.

(5) $\bigcap_{n\in\mathbb{N}} A_n = [0,1)$ and $\bigcup_{n\in\mathbb{N}} A_n = (-1,1)$.

SOLUTION 1.2.8. Let $a, b \geq 0$ and let $p, q > 1$. If $a = 0$ or $b = 0$, then it is evident that Young's inequality is verified. So, assume that $a, b > 0$ and let $f : \mathbb{R}^+ \to \mathbb{R}$ be the function defined by

$$f(x) = \frac{x^p}{p} + \frac{1}{q} - x.$$

It can easily be established that $f(x) \geq 0$ for all $x \geq 0$. In particular, for $x = ab^{\frac{1}{1-p}}$ we have

$$ab^{\frac{1}{1-p}} \leq \frac{(ab^{\frac{1}{1-p}})^p}{p} + \frac{1}{q} = \frac{a^p b^{\frac{p}{1-p}}}{p} + \frac{1}{q}$$

and hence

$$ab^{\frac{1}{1-p}} b^q \leq \frac{a^p b^{\frac{p}{1-p}} b^q}{p} + \frac{b^q}{q}.$$

But $\frac{1}{p} + \frac{1}{q} = 1$ implies that

$$b^{\frac{1}{1-p}} b^q = b \quad \text{and} \quad b^{\frac{p}{1-p}} b^q = 1.$$

In the end, we obtain

$$ab \leq \frac{a^p}{p} + \frac{b^q}{q},$$

completing the proof.

SOLUTION 1.2.9. The sequence $(x_n)_n$ is strictly increasing and the sequence $(y_n)_n$ is strictly decreasing and it is well-known that

$$\lim_{n\to\infty} x_n = \lim_{n\to\infty} y_n = e$$

and we have $x_n < e < y_n$. Now, assume that $e = \frac{a}{b}$ where $a \in \mathbb{N}$ and $b \in \mathbb{N}$. Since $x_n < e < y_n$ for all n, taking $n = b$ gives us

$$x_b < e = \frac{a}{b} < y_b, \quad \text{i.e. } x_b < e = \frac{a}{b} < x_b + \frac{1}{b!b}.$$

Hence we are led to

$$b!bx_b < b!a < b!bx_b + 1.$$

But $b!bx_b$ is an integer which we denote by p. Hence we have $p < b!a < p+1$ (where $p \in \mathbb{N}$) which is impossible since $b!a \in \mathbb{N}$. Thus $e \in \mathbb{R} \setminus \mathbb{Q}$.

REMARK. In fact, e is not even an algebraic number (the the interested reader may consult any standard book in number theory).

SOLUTION 1.2.10. Let $x, y \in \mathbb{R}$. We may assume WLOG that $x < y$. We are required to find an $r \in \mathbb{Q}$ such that $x < r < y$. Set $a = y - x > 0$. By the Archimedean property, for some $p \in \mathbb{N}$, $pa > 1$ or $\frac{1}{p} < a$. Set $q = [px] + 1$ where $[\cdot]$ denotes the greatest integer function. Then $q - 1 \leq px < q$. Hence

$$x < \frac{q}{p} \leq x + \frac{1}{p} < x + a = y$$

and the proof will be complete by taking $r = \frac{q}{p}$.

SOLUTION 1.2.11. To prove the inequality, we first note that

$$\forall x \in [0, 1] : \ |f(x)| \leq \sup_{0 \leq x \leq 1} |f(x)|$$

and hence

$$\forall x \in [0, 1] : \ |f(x)||g(x)| \leq \sup_{0 \leq x \leq 1} |f(x)||g(x)|.$$

Integrating with respect to x on $[0, 1]$ and taking into account that the final value of $\sup_{0 \leq x \leq 1} |f(x)|$ is a number.

$$\int_0^1 |f(x)g(x)|dx = \int_0^1 |f(x)||g(x)|dx \leq \int_0^1 \sup_{0 \leq x \leq 1} |f(x)||g(x)|dx$$

$$= \sup_{0 \leq x \leq 1} |f(x)| \int_0^1 |g(x)|dx.$$

CHAPTER 2

Metric Spaces

2.2. True or False: Answers

ANSWERS.

(1) Yes, an instance of that is the discrete metric (see Exercise 2.3.5).

(2) The answer is yes. For a proof, see Test 1.

(3) In topology, especially for beginners, this is something to avoid absolutely. We always have to emphasize with respect to which metric space we are working. Knowing this, we may consider this question as false or badly formulated. The given set is not open in $\mathbb{R}$ endowed with the usual metric. But with respect to the discrete metric, it becomes open as every subset is open in this metric space (see Exercise 2.3.5 below).

(4) No! It is not open, but certainly not because it is closed! It is not open for

$$\forall r > 0, \ (-r, r) \not\subset \{0\}.$$

(5) The answer is yes! This may come as a surprise to some of the readers but they have to be prepared to more "surprises" (interesting though) in Topology. We always remind the students that not every thing true in the classical $\mathbb{R}$, $\mathbb{C}$, $\mathbb{R}^n$,...etc will be true in an arbitrary metric or topological space.

   Let us go back to our question. We give two counterexamples for the sake of diversity of examples.

   (a) Let $\mathbb{R}$ be endowed with the *discrete metric* (see Exercise 2.3.5 below). Then

$$B\left(0, \frac{2}{3}\right) = \{0\} \subseteq B\left(0, \frac{1}{2}\right) = \{0\}$$

   (this also shows that two balls of the same center with different radii may well be identical!).

   (b) Let $\mathbb{N}$ be endowed with the induced standard metric of $\mathbb{R}$. Then

$$B_c(3, 2) = \{n \in \mathbb{N} : \ |n - 3| \leq 2\} = \{1, 2, 3, 4, 5\}$$

and

$$B_c(0,4) = \{n \in \mathbb{N} : |n| \le 4\} = \{1,2,3,4\} \subsetneq B_c(3,2)$$

(6) The answer is no in general. We will show in Exercise 2.3.3 that

$$d(x,y) = \left| \frac{1}{x} - \frac{1}{y} \right|$$

defines a metric on $\mathbb{R}^*$. The given set, i.e. $(0,1)$ is not bounded with respect to this metric since $\left| \frac{1}{x} - \frac{1}{y} \right| \to +\infty$ as $x, y \to 0^+$.

REMARK. With respect to this metric, $\mathbb{N}$ or more generally, $[1,\infty)$ (which are obviously not bounded in usual $\mathbb{R}$) are bounded sets!

(7) The answer is no. In $(\mathbb{R}, |\cdot|)$ ($\mathbb{R}$ endowed with the usual metric), consider

$$A = [-1,1], \ B = [1,2] \text{ and } C = [2,4].$$

Then

$$d(A,C) = 1 > d(A,B) + d(B,C) = 0.$$

(8) The proof is easy and we shall prove only one implication. The other implication, being very akin, is left to the interested reader. Let U be an open set in (X,d). Then, for each $x \in U$, we can find some $r > 0$ such that $B(x,r) \subset U$. Since d and d' are equivalent, there are some $\alpha, \beta > 0$ such that for all $x, y \in X$ one has

$$\alpha d'(x,y) \le d(x,y) \le \beta d'(x,y).$$

Then if $B'(x,r')$ is the open ball in (X,d'), where $r' = \frac{r}{\beta}$, and if $y \in B'(x,r')$, then

$$d(x,y) \le \beta d'(x,y) < \beta \frac{r}{\beta} = r,$$

implying that $y \in B(x,r) \subset U$ and hence $B'(x,r') \subset U$. Hence U is open in (X',d').

(9) The answer is yes. Let (X,d) be a metric space. Let $B_c(x,r)$ be the closed ball of radius $r > 0$ and center x. We need only prove that the complement of $B_c(x,r)$ is open in (X,d). We note that

$$B_c(x,r)^c = \{y \in X : d(x,y) > r\}.$$

Take $s = d(x, y) - r > 0$. Let $z \in B(x, s)$, i.e. $d(x, z) < s = d(x, y) - r$ and hence

$$r < d(x, y) - d(x, z) \leq d(x, z) + d(z, y) - d(x, z) = d(z, y),$$

i.e. $B(x, s) \subset B_c^c(x, r)$ proving the openness of $B_c^c(x, r)$ or the closedness of $B_c(x, r)$.

The converse is not true in general. Consider the same idea for a counterexample as the one in Exercise 2.3.17, where it is also proved that every open ball in a metric space is an open set.

(10) We recall that, in a metric space (X, d), the sphere of center x and radius $r > 0$ is given by

$$S(x, r) = \{y \in X : d(x, y) = r\}$$

which can be written as

$$S(x, r) = B_c(x, r) \cap [B(x, r)]^c,$$

i.e. as a *finite* intersection of closed sets and hence $S(x, r)$ is closed (see Exercise 2.5.6).

(11) The answer is yes and this happens for example in a ultrametric space (see Exercise 2.3.20).

2.3. Solutions to Exercises

SOLUTION 2.3.1. We need only show the triangle inequality for x, y and z not all distinct. If $x = z$, then $d(x, z) = 0$ and hence

$$0 = d(x, z) \leq d(x, y) + d(y, z).$$

If $x = y$, then

$$d(x, z) = d(y, z) = d(x, y) + d(y, z),$$

i.e. we have equality in this case and so we will have in the case $y = z$. The proof is over.

REMARK. This result may be helpful for proving a given function is a metric especially when there are different cases to look at. See for instance the proof for the discrete metric below.

SOLUTION 2.3.2. Let $x, y, z \in X$. We have

$$d(x, z) \leq d(x, y) + d(y, z) \text{ and hence } d(x, z) - d(y, z) \leq d(x, y).$$

Inverting the roles of x and y yields

$$d(y, z) - d(x, z) \leq d(y, x) = d(x, y).$$

Thus

$$|d(x, z) - d(y, z)| \leq d(x, y).$$

SOLUTION 2.3.3.

(1) No, d is not a metric. For example, $1 \neq -1$ but $d(1, -1) = 0$.

(2) Yes, d is a metric on $\mathbb{R}$. Let us show that. First, we note that the range of d is $\mathbb{R}^+$. Let us check the remaining properties of a metric.

(a) Let $x, y \in \mathbb{R}$. Then
$$d(x, y) = 0 \Leftrightarrow |x^3 - y^3| = 0 \Leftrightarrow x^3 = y^3 \Leftrightarrow x = y.$$

(b) Let $x, y \in \mathbb{R}$. Then
$$d(x, y) = |x^3 - y^3| = |y^3 - x^3| = d(y, x).$$

(c) Let $x, y, z \in \mathbb{R}$. Then
$$d(x, z) = |x^3 - z^3| = |x^3 - y^3 + y^3 - z^3| \leq |x^3 - y^3| + |y^3 - z^3|$$

and hence
$$d(x, z) \leq d(x, y) + d(y, z).$$

(3) No, since d never vanishes.

(4) Yes, d is indeed a metric. Let us show this.

(a) Let $x, y \in \mathbb{R}^*$. We have
$$x = y \Longleftrightarrow \frac{1}{x} = \frac{1}{y} \Longleftrightarrow \left|\frac{1}{x} - \frac{1}{y}\right| = 0 \Longleftrightarrow d(x, y) = 0.$$

(b) Let $x, y \in \mathbb{R}^*$. We obviously have
$$d(x, y) = d(y, x).$$

(c) Let $x, y, z \in \mathbb{R}^*$. We have
$$d(x, z) = \left|\frac{1}{x} - \frac{1}{z}\right| = \left|\frac{1}{x} - \frac{1}{y} + \frac{1}{y} - \frac{1}{z}\right| \leq \left|\frac{1}{x} - \frac{1}{y}\right| + \left|\frac{1}{y} - \frac{1}{z}\right|$$

and so
$$d(x, z) \leq d(x, y) + d(y, z).$$

(5) No, since $d(1, 1) = 2 \neq 0$.

SOLUTION 2.3.4. The open ball of center x and radius $r > 0$ in the usual metric of $\mathbb{R}$ is given by
$$B(x, r) = \{y \in \mathbb{R} : |x - y| < r\} = (x - r, x + r).$$
So if $(x - r, x + r) = (0, 1)$, then
$$\begin{cases} x - r = 0, \\ x + r = 1 \end{cases} \text{ and whence } \begin{cases} x = \frac{1}{2}, \\ r = \frac{1}{2}. \end{cases}$$

SOLUTION 2.3.5.

(1) Clearly $\forall x, y \in X$, $d(x,y) \geq 0$. Also, the first two properties of a metric are evidently satisfied.

Now, let us show the triangle inequality. If $x = z$, then $d(x,z) = 0$ and hence

$$d(x,z) \leq d(x,y) + d(y,z).$$

If $x \neq z$, then either $x \neq y$ or $y \neq z$ (otherwise we will have $x = y = z$) and hence

$$1 = d(x,z) \leq d(x,y) + d(y,z) = 1 \text{ or } 2.$$

(2) We show that

$$B(x,r) = \{x\} \text{ if } r \leq 1 \text{ and } B(x,r) = X \text{ if } r > 1.$$

We have

$$B(x,r) = \{y \in X : \ d(x,y) < r \leq 1\} \subset \{y \in X : \ d(x,y) < 1\} = \{x\}$$

since $d(x,y)$ is worth 1 or 0. On the other hand, it is plain that $\{x\} \subset B(x,r)$, hence the equality holds. Since

$$d(x,y) \leq 1 < r, \ \forall x, y \in X,$$

we see that $B(x,r) = X$.

Similarly, we can prove that

$$B_c(x,r) = \{x\} \text{ for } r < 1 \text{ and } B_c(x,r) = X \text{ for } r \geq 1.$$

(3) If $r \neq 1$ (remember that $r > 0$), then obviously $S(x,r) = \varnothing$. If, however, $r = 1$, then

$$S(x,1) = \{y \in X : \ d(x,y) = 1\} = X \setminus \{x\}.$$

(4) The way of proving $\varnothing$ and X are open is the same as in Exercise 2.3.15 below. Now, let U be any subset of X. Let $x \in U$. Then

$$B\left(x, \frac{1}{2}\right) = \{x\} \subset U,$$

proving the openness of U.

(5) If V is a subset of X, then so is its complement V^c. Hence V^c is open and thus V is closed.

(6) Denote the usual metric on $\mathbb{R}$ by $|\cdot|$. To show d is not equivalent to $|\cdot|$, we can equivalently show that the function $(x,y) \mapsto \frac{d(x,y)}{|x-y|}$ is not bounded on $\mathbb{R}^2$. If it were, then it would have been so for $y = 0$ and $x = e^{-n}$. But

$$\frac{d(e^{-n}, 0)}{|e^{-n}|} = \frac{1}{e^{-n}} = e^n \to \infty \text{ as } n \to \infty.$$

Thus the two metrics are not equivalent.

SOLUTION 2.3.6. We first note that d' is a positive and well-defined function on $X \times X$ since d is a metric (the fact that d is a metric will be used a few times and it will be clear to the reader without further notice). Now we verify the other axioms.

(1) Let $x, y \in X$.
    (a) If $x = y$, then $d(x, y) = 0$. Hence $d'(x, y) = \sqrt{d(x, y)} = 0$.
    (b) If $d'(x, y) = 0$, then $\sqrt{d(x, y)} = 0$ or $d(x, y) = 0$. Thus $x = y$.
(2) Let $x, y \in X$. We obviously have
$$d'(x, y) = \sqrt{d(x, y)} = \sqrt{d(y, x)} = d'(y, x).$$
(3) Let $x, y, z \in X$. Since $d(x, z) \leq d(x, y) + d(y, z)$ (as d is a metric), since $\sqrt{\cdot}$ is increasing on $\mathbb{R}^+$ and since $\sqrt{a + b} \leq \sqrt{a} + \sqrt{b}$ (for all $a, b \geq 0$), one can write
$$d'(x, z) = \sqrt{d(x, z)} \leq \sqrt{d(x, y) + d(y, z)} \leq \sqrt{d(x, y)} + \sqrt{d(y, z)}$$
and hence
$$d'(x, z) \leq d'(x, y) + d'(y, z).$$

SOLUTION 2.3.7.

(1) Let $x, y \in \mathbb{N}$, then obviously $d(x, y) = 0 \Leftrightarrow x = y$.
(2) Also, for all $x, y \in \mathbb{N}$, $d(x, y) = d(y, x)$.
(3) Lastly, let $x, y, z \in \mathbb{N}$. Then
$$3 + \frac{x + z}{xz} = 3 + \frac{1}{x} + \frac{1}{z} \leq 3 + \frac{1}{x} + \frac{1}{y} + 3 + \frac{1}{y} + \frac{1}{z}$$
and hence
$$d(x, z) \leq d(x, y) + d(y, z), \ \forall x, y, z \in \mathbb{N}.$$

SOLUTION 2.3.8. First, observe that for all $x, y \in \mathbb{R}^n$, $d(x, y) \geq 0$. Let us show now the axioms of a metric.

(1) If $x = y$, then $|x_k - y_k| = 0$ (for all $k = 1, \cdots, n$) and hence $d(x, y) = 0$. Conversely, let $x, y \in \mathbb{R}^n$. We have
$$d(x, y) = 0 \Rightarrow |x_k - y_k| = 0, \ \forall k \in \{1, \cdots, n\} \Rightarrow x_k = y_k, \ \forall k \in \{1, \cdots, n\},$$
i.e. $x = y$.
(2) Let $x, y \in \mathbb{R}^n$. We have
$$d(x, y) = \sum_{k=1}^{n} |x_k - y_k| = \sum_{k=1}^{n} |y_k - x_k| = d(y, x).$$

(3) Let $x, y, z \in \mathbb{R}^n$. Since

$$|x_k - z_k| \leq |x_k - y_k| + |y_k - z_k|, \ \forall k \in \{1, \cdots, n\},$$

summing in k yields

$$d(x, z) \leq d(x, y) + d(y, z).$$

SOLUTION 2.3.9.

(1) Set for any $1 \leq k \leq n$,

$$A_k = \frac{a_k}{\left(\sum_{k=1}^n a_k^p\right)^{\frac{1}{p}}} \text{ and } B_k = \frac{b_k}{\left(\sum_{k=1}^n b_k^q\right)^{\frac{1}{q}}}.$$

Applying Young's inequality (see Exercise 1.2.8) to A_k and B_k gives us $A_k B_k \leq \frac{A_k^p}{p} + \frac{B_k^q}{q}$, i.e.

$$\frac{a_k}{\left(\sum_{k=1}^n a_k^p\right)^{\frac{1}{p}}} \frac{b_k}{\left(\sum_{k=1}^n b_k^q\right)^{\frac{1}{q}}} \leq \frac{1}{p} \frac{a_k^p}{\sum_{k=1}^n a_k^p} + \frac{1}{q} \frac{b_k^q}{\sum_{k=1}^n b_k^q}.$$

Summing in k yields

$$\frac{\sum_{k=1}^n a_k b_k}{\left(\sum_{k=1}^n a_k^p\right)^{\frac{1}{p}} \left(\sum_{k=1}^n b_k^q\right)^{\frac{1}{q}}} \leq \frac{1}{p} + \frac{1}{q} = 1$$

and the desired result will then follows easily.

(2) To prove the Minkowski inequality, we will use the Hölder's inequality twice. We have

$$\sum_{k=1}^n (a_k + b_k)^p = \sum_{k=1}^n (a_k + b_k)^{p-1}(a_k + b_k)$$

$$= \sum_{k=1}^n (a_k + b_k)^{p-1} a_k + \sum_{k=1}^n (a_k + b_k)^{p-1} b_k$$

$$\leq \left(\sum_{k=1}^n (a_k + b_k)^{(p-1)q}\right)^{\frac{1}{q}} \left(\sum_{k=1}^n a_k^p\right)^{\frac{1}{p}}$$

$$+ \left(\sum_{k=1}^n (a_k + b_k)^{(p-1)q}\right)^{\frac{1}{q}} \left(\sum_{k=1}^n b_k^p\right)^{\frac{1}{p}}.$$

But $\frac{1}{p} + \frac{1}{q} = 1 \Rightarrow (p-1)q = p$ and so

$$\sum_{k=1}^{n}(a_k + b_k)^p \leq \left(\sum_{k=1}^{n}(a_k + b_k)^p\right)^{\frac{1}{q}}\left(\sum_{k=1}^{n}a_k^p\right)^{\frac{1}{p}}$$

$$+ \left(\sum_{k=1}^{n}(a_k + b_k)^p\right)^{\frac{1}{q}}\left(\sum_{k=1}^{n}b_k^p\right)^{\frac{1}{p}}$$

$$= \left(\sum_{k=1}^{n}(a_k + b_k)^p\right)^{\frac{1}{q}}\left(\left(\sum_{k=1}^{n}b_k^p\right)^{\frac{1}{p}} + \left(\sum_{k=1}^{n}a_k^p\right)^{\frac{1}{p}}\right).$$

Hence (and since $\frac{1}{p} = 1 - \frac{1}{q}$)

$$\left(\sum_{k=1}^{n}(a_k + b_k)^p\right)^{1-\frac{1}{q}} = \left(\sum_{k=1}^{n}(a_k + b_k)^p\right)^{\frac{1}{p}} \leq \left(\sum_{k=1}^{n}b_k^p\right)^{\frac{1}{p}} + \left(\sum_{k=1}^{n}a_k^p\right)^{\frac{1}{p}}$$

and the proof is over.

(3) (a) We first prove that d_∞ is a metric and this does not require the previous questions.

(i) Let $x, y \in \mathbb{R}^n$. If $x = y$, then obviously $d_\infty(x, y) = 0$.

Now, if $d_\infty(x, y) = 0$, then for all $1 \leq k \leq n : |x_k - y_k| = 0$ leading to $x = y$.

(ii) We evidently have for all $x, y \in \mathbb{R}^n$

$$d_\infty(x, y) = d_\infty(y, x)$$

(iii) Let $x, y, z \in \mathbb{R}^n$. Then for all $1 \leq k \leq n$, we have

$$|x_k - z_k| \leq |x_k - y_k| + |y_k - z_k| \leq d_\infty(x, y) + d_\infty(y, z)$$

and hence

$$d_\infty(x, z) \leq d_\infty(x, y) + d_\infty(y, z).$$

Now we show that each d_p is a metric. Let $p \geq 1$.

(i) Let $x, y \in \mathbb{R}^n$. Then if $d_p(x, y) = 0$, one will have

$$\sum_{k=1}^{n}|x_k - y_k|^p = 0 \Rightarrow |x_k - y_k|^p = 0, \ \forall k = 1, \cdots, n \Rightarrow x = y.$$

The other implication is obvious.

(ii) The second axiom is verified as for all $x, y \in \mathbb{R}^n$

$$d_p(x, y) = d_p(y, x).$$

(iii) The triangle inequality is a mere consequence of Minkowski inequality. Setting and reporting $x_k - y_k = a_k$ and $y_k - z_k = b_k$ (for all $k = 1, \cdots, n$) into Minkowski inequality give us the desired triangle inequality.

(b) First we show the equivalence of the metrics. We have for all $1 \leq k \leq n$,

$$|x_k - y_k|^p \leq \sum_{k=1}^{n} |x_k - y_k|^p = d_p^p(x, y).$$

Hence, taking the pth root (without changing the sign of the inequality which is legitimate as the function $x \mapsto x^a$, for $a > 0$, is increasing on $\mathbb{R}_+^*$) and taking the max over k gives

$$d_\infty(x, y) \leq d_p(x, y)$$

and we are half-way through. We also have for all $1 \leq k \leq n$,

$$|x_k - y_k|^p \leq \max_{1 \leq k \leq n} (|x_k - y_k|^p) = d_\infty^p(x, y)$$

and summing in k and the taking the pth root yield

$$d_p(x, y) \leq n^{\frac{1}{p}} d_\infty(x, y),$$

demonstrating the equivalence of the metrics.
To prove the last desired limit, we note that since for all $p \geq 1$,

$$d_\infty(x, y) \leq d_p(x, y) \leq n^{\frac{1}{p}} d_\infty(x, y),$$

taking the limit as $p \to \infty$ and observing that $n^{\frac{1}{p}} \to 1$, we easily obtain

$$\lim_{p \to \infty} d_p(x, y) = d_\infty(x, y).$$

SOLUTION 2.3.10.

(1) (a) Let us show that δ is in effect a metric on X.

    (i) Let $x, y \in X$. We have

$$\delta(x, y) = 0 \iff \min(1, d(x, y)) = 0 \iff d(x, y) = 0 \iff x = y$$

since d is a distance.

    (ii) Let $x, y \in X$. It is plain that $\delta(x, y) = \delta(y, x)$.

    (iii) Let $x, y, z \in X$.

        (A) If $d(x, y) \leq 1$ and $d(y, z) \leq 1$, then

$$\delta(x, z) \leq d(x, z) \leq d(x, y) + d(y, z) = \delta(x, y) + \delta(y, z).$$

(B) If $d(x, y) > 1$, then

$$\delta(x, z) \leq 1 = \delta(x, y) \leq \delta(x, y) + \delta(y, z).$$

(C) The same arguments apply if $d(y, z) > 1$.

(b) Now we show that ρ is a metric on X.

   (i) The first property of a metric is trivial.

   (ii) The second property is also trivial.

   (iii) Now, let $x, y, z \in \mathbb{R}$. Since $d(x, z) \leq d(x, y) + d(y, z)$ and using the hint, we easily see that

$$\rho(x, z) = \frac{d(x, z)}{1 + d(x, z)} \leq \frac{d(x, y) + d(y, z)}{1 + d(x, y) + d(y, z)} \leq \frac{d(x, y)}{1 + d(x, y)} + \frac{d(y, z)}{1 + d(y, z)}.$$

Thus ρ is a metric on X.

(2) The set X is bounded (even if X is $\mathbb{R}$, say) with respect to both metrics since the given metrics are bounded as

$$\forall x, y \in X : \ \delta(x, y) \leq 1 \text{ and } \rho(x, y) \leq 1.$$

REMARK. By doing some arithmetic one can prove the triangle inequality for ρ directly without calling on the hint (the reader should try it).

SOLUTION 2.3.11. To show that d is a metric, we first need to check that the series involved in the definition of d converges. For any n, one always have

$$\frac{1}{2^n} \times \frac{d_n(x_n, y_n)}{1 + d_n(x_n, y_n)} \leq \frac{1}{2^n} = u_n.$$

Since $\sum_{n=1}^{\infty} u_n = 1 < +\infty$, $d(x, y) < \infty$.

Now the previous exercise can be used to show the properties of a metric for d and we leave the details to the reader.

SOLUTION 2.3.12.

(1) Since f and g are continuous, so is $|f - g|$ and hence $d(f, g)$ is a well-defined quantity and so is $d'(f, g)$ as $|f - g|$ is continuous on the *closed* and *bounded* $[0, 1]$. Observe that both d and d' have $\mathbb{R}^+$ as their range.

   First, we show that d is a metric on X.

(a) Let $f, g \in X$. We have

$$f = g \Longrightarrow d(f, g) = 0 \text{ (this is obvious)}.$$

Now since $|f - g|$ is a *continuous* and *positive* function, we have

$$\int_0^1 |f(x) - g(x)| dx = 0 \implies |f(x) - g(x)| = 0, \ \forall x \in [0, 1] \iff f = g.$$

(b) It is plain that for all f and g in X one has $d(f, g) = d(g, f)$.

(c) Let us show now the triangle inequality. Let $f, g, h \in X$. We have for all $x \in [0, 1]$

$$|f(x) - h(x)| = |f(x) - g(x) + g(x) - h(x)| \le |f(x) - g(x)| + |g(x) - h(x)|.$$

Hence

$$\int_0^1 |f(x) - h(x)| dx \le \int_0^1 |f(x) - g(x)| dx + \int_0^1 |g(x) - h(x)| dx,$$

i.e. $d(f, h) \le d(f, g) + d(g, h)$.

Now we prove that d' is a metric on X too.

(a) If $f = g$, then obviously $d'(f, g)$, being the supremum of the nil function, must vanish as well. Conversely, if $d'(f, g) = 0$, then for all $x \in [0, 1]$ we have

$$0 \le |f(x) - g(x)| \le \sup_{x \in [0,1]} |f(x) - g(x)| = d'(f, g) = 0.$$

Thus $f = g$.

(b) For all $f, g \in X$, we have

$$d'(f, g) = \sup_{x \in [0,1]} |f(x) - g(x)| = \sup_{x \in [0,1]} |g(x) - f(x)| = d'(g, f).$$

(c) Let $f, g, h \in X$. Then we have for all $x \in [0, 1]$

$$|f(x) - h(x)| \le |f(x) - g(x)| + |g(x) - h(x)| \le d'(f, g) + d'(g, h).$$

Thus

$$d'(f, h) \le d'(f, g) + d'(g, h).$$

(2) We recall that d and d' are said to be equivalent if

$$\exists a, b > 0, \ \forall f, g \in X : \ ad'(f, g) \le d(f, g) \le bd'(f, g).$$

We show that the LHS inequality does not hold (the RHS one does hold, cf Exercise 1.2.11) and in order to simplify the proof a bit we may take $g = 0$. Now take $f(x) = x^n$, defined on $[0, 1]$, for a given n in $\mathbb{N}$. Then $d(f, 0) = \frac{1}{n+1}$ and $d'(f, 0) = 1$. If d were equivalent to d' we would have

$$\exists a > 0, \forall n \in \mathbb{N} : \ a \le \frac{1}{n + 1}$$

which would imply that $\mathbb{N}$ is bounded, something impossible, i.e. we reached a contradiction. Thus d and d' are not equivalent.

(3) No, d is no longer a metric in this case. For if one takes

$$f(x) = \begin{cases} 0, & x \in [0,1), \\ 2, & x = 1. \end{cases}$$

and g is the zero function, then $f \neq g$ and yet

$$d(f,g) = \int_0^1 |f(x) - 0| dx = 0.$$

SOLUTION 2.3.13. The answer is no. The function d does verify the second and third property of a metric and one implication of the first one (i.e. $f = g \Rightarrow d(f,g) = 0$). However,

$$d(f',g') = 0 \nRightarrow f = g.$$

For instance take any $f \in X$ and $g(x) = f(x) + 1$. Then $f \neq g$ but $d(f,g) = 0$.

SOLUTION 2.3.14.

(1) The answer is no. On simple instance is to take $\mathbb{R}$ equipped with the discrete metric (see Exercise 2.3.5). Take $r = 2$ and $s = 3$ and $x = 0$ and $y = -1$. Then we know that $B(0,2) = B(-1,3) = \mathbb{R}$ and of course $r \neq s$ and $x \neq y$.

(2) The result is true for example in $\mathbb{R}$ endowed with its standard metric.

SOLUTION 2.3.15.

(1) First, we show that X is open. Let $x \in X$. Then for some (and in this case all) $r > 0$

$$B(x,r) \subset X.$$

Thus X is open in (X, d). The openness of $\varnothing$ follows from the observation that if $x \in \varnothing$ (x does not exist!), then for some (and here any) $r > 0$

$$B(x,r) = \varnothing \subset \varnothing$$

(2) Let $\{U_i\}_{i \in I}$ be an arbitrary family of open sets in (X, d). We have to show that their union remains open. To this end, let $x \in \bigcup_{i \in I} U_i$. Then there is some $j \in I$ such that $x \in U_j$. But,

U_j is open and hence it contains an open ball, $B(x, r)$ $(r > 0)$ say. Hence

$$B(x, r) \subset U_j \subset \bigcup_{i \in I} U_i.$$

Thus the arbitrary union of open sets is open.

(3) We only prove it for two open sets. The proof for a finite intersection follows easily by induction. Let U and V be two open sets in X. Let $x \in U \cap V$. Hence $x \in U$ and $x \in V$. Then

$$\exists r > 0, \ B(x, r) \subset U \text{ and } \exists s > 0, \ B(x, s) \subset V$$

Taking $a = \min(r, s)$ leads to

$$B(x, a) \subset U \cap V,$$

showing the openness of $U \cap V$.

The last result does not extend to infinite intersections. For instance, all the intervals $(-\frac{1}{n}, \frac{1}{n})$ are open (for all $n \in \mathbb{N}$), however,

$$\bigcap_{n \in \mathbb{N}} \left(-\frac{1}{n}, \frac{1}{n} \right) = \{0\}$$

which is not open.

SOLUTION 2.3.16. Assume that a set U is written as a union of open balls. Then, obviously U is open since every open ball is an open set and the arbitrary union of open sets is open (see the foregoing exercise!).

Conversely, suppose U is open. By definition of an open set in a metric space, for every x in U we can always find $r > 0$ for which $B(x, r) \subset U$. Whence

$$U = \bigcup_{x \in U} \{x\} \subset \bigcup_{x \in U} B(x, r) \subset U, \text{ i.e. } U = \bigcup_{x \in U} B(x, r).$$

SOLUTION 2.3.17. Let (X, d) be a metric space and let $B(X, r)$ be an open ball with center $x \in X$ and radius $r > 0$. Let $y \in B(x, r)$ (so $d(x, y) < r$). We must show that $B(x, r)$ contains another open ball centered at y. Let $s = r - d(x, y) > 0$. Then it may be easily showed that

$$B(y, s) \subset B(x, r),$$

completing the proof.

There are many counterexamples for the converse. For instance, in a discrete metric space X (for convenience choose X such that $\operatorname{card} X \geq 3$), open balls are only X or singletons (nothing else). On the other hand, it is known that any subset is open, so $\{x, y\}$, where $x, y \in X$, is open too but as observed just above, it is not an open ball.

SOLUTION 2.3.18. None of the given intervals is open in $\mathbb{R}$. For instance $[a, b)$ is not open in $\mathbb{R}$ since $a \in [a, b)$ and

$$\forall r > 0, B(a, r) = (a - r, a + r) \not\subset [a, b).$$

SOLUTION 2.3.19.

(1) Straightforward verification for the first two properties. As for the triangle inequality, let $x, y, z \in \mathbb{R}^2$. Then we have

$$\begin{aligned}
\delta(x, z) &= d(x, \mathbf{0}) + d(z, \mathbf{0}) \\
&\leq d(x, \mathbf{0}) + d(y, \mathbf{0}) + d(y, \mathbf{0}) + d(z, \mathbf{0}) \\
&= \delta(x, y) + \delta(y, z).
\end{aligned}$$

(2) Let $a \neq \mathbf{0}$. We have

$$B(a, r) = \{x \in \mathbb{R}^2 : \ d(x, \mathbf{0}) + d(a, \mathbf{0}) < r\}$$

if $x \neq a$. Choosing $r < d(a, \mathbf{0})$, e.g. $r = \frac{1}{3}d(a, \mathbf{0})$ (which is legitimate since $a \neq \mathbf{0}$), we see that $d(x, \mathbf{0}) + d(a, \mathbf{0}) < r$ cannot be realized and hence we are left with $\delta(x, a) = 0$, i.e. $B(a, \frac{1}{3}d(a, \mathbf{0})) = \{a\}$.

(3) If $\{\mathbf{0}\}$ were open in $(\mathbb{R}^2, \delta)$, then there would exist a positive r such that $B(\mathbf{0}, r) \subset \{\mathbf{0}\}$. But

$$B(\mathbf{0}, r) = \{(x, y) \in \mathbb{R}^2 : \ \delta((x, y), \mathbf{0}) < r\} = \{(x, y) \in \mathbb{R}^2 : \ x^2 + y^2 < r^2\}$$

cannot be a subset of $\{\mathbf{0}\}$ for every $r > 0$. Thus $\{\mathbf{0}\}$ is not open in $(\mathbb{R}^2, \delta)$.

(4) In $\mathbb{R}^2 \setminus \{\mathbf{0}\}$ every singleton is open. Hence every set is open since it can be written as a union (even arbitrary, see Exercise 2.3.16) of singletons. Thus every set in $\mathbb{R}^2 \setminus \{\mathbf{0}\}$ is closed too. Accordingly, every subset of $\mathbb{R}^2 \setminus \{\mathbf{0}\}$ is clopen and hence δ coincides with the discrete metric on $\mathbb{R}^2 \setminus \{\mathbf{0}\}$.

SOLUTION 2.3.20.

(1) The only property to prove is the triangle inequality. We may assume WLOG that $d(x, y) < d(y, z)$. Then

$$d(x, z) \leq \max(d(x, y), d(y, z)) = d(y, z) \leq d(x, y) + d(y, z).$$

(2) Assume $d(x, y) \neq d(y, z)$ or we could just assume that $d(x, y) < d(y, z)$ (couldn't we?). We can write by hypothesis

$$(\ d(x, y) <)\ d(y, z) \leq \max(d(x, y), d(x, z))$$

and a fortiori

$$\max(d(x,y), d(x,z)) = d(x,z), \text{i.e. } d(y,z) \leq d(x,z).$$

The "ultrametricity" hypothesis also gives us

$$d(x,z) \leq \max(d(x,y), d(y,z)) = d(y,z)$$

and so $d(x,z) = d(y,z)$.

Geometrically, this means that in a ultrametric space every triangle is isosceles.

(3) Let $B(x,r)$ be the open ball of center x and of radius $r > 0$. Let $y \in B(x,r)$ (hence $d(x,y) < r$). We need to show that $B(x,r) = B(y,r)$. Let $z \in B(x,r)$, i.e. $d(x,z) < r$. Hence

$$d(y,z) \leq \max(d(x,y), d(x,z)) < r \Rightarrow z \in B(y,r).$$

The other inclusion can be dealt with similarly.

(4) Well, every closed ball is a closed set *in any metric space*. We show that every closed ball is open too. Let $B_c(x,r)$ be a closed ball of center x and of radius $r > 0$. Let $y \in B_c(x,r)$, i.e. $d(x,y) \leq r$. We are done if we show that this closed ball contains the open ball $B(y,r)$. Let $z \in B(y,r)$. Then

$$d(x,z) \leq \max(d(x,y), d(y,z)) \leq r$$

and thus $z \in B_c(x,r)$.

We leave to the reader to show that every open ball is closed.

SOLUTION 2.3.21. First, note that the definition which was recalled can be re-written as follows: f is continuous at x iff

$$\forall \varepsilon > 0, \exists \delta > 0, \forall y \in X : (d(x,y) < \delta \Rightarrow d'(f(x), f(y)) < \varepsilon).$$

Let us go back now to the proof. Assume that f is continuous and let U be an open set in X'. We ought to show that $f^{-1}(U)$ is open in X. Let x be in $f^{-1}(U)$, i.e. $f(x) \in U$. But U is open and hence there exists some $\varepsilon > 0$, $B(f(x), \varepsilon) \subset U$. Since f is continuous, the hypothesis implies that

$$f(B(x,\delta)) \subset B(f(x), \varepsilon) \subset U \text{ which leads to } B(x,\delta) \subset f^{-1}(U),$$

proving the openness of $f^{-1}(U)$.

Conversely, let us show that f is continuous at $x \in X$ (assuming that the preimage by f of every open set in X' is open in X). Let $\varepsilon > 0$. Then it is clear that $B(f(x), \varepsilon)$ is open in X'. Then by hypothesis, $f^{-1}(B(f(x), \varepsilon))$ is also an open set but in X. This guarantees the existence of a strictly positive δ such that

$$B(x,\delta) \subset f^{-1}(B(f(x), \varepsilon))$$

or

$$f(B(x,\delta)) \subset B(f(x),\varepsilon),$$

establishing the continuity of f.

SOLUTION 2.3.22.

(1) Let $f : (X, d) \to (\mathbb{R}, |\cdot|)$ be a function such that $f(x) = d(x, a)$. Then f is continuous on X (it is in fact uniformly continuous) because for any $x, y \in X$

$$|f(x) - f(y)| = |d(x, a) - d(y, a)| \le d(x, y) \text{ (by Exercise 2.3.2)}.$$

(2) Let x, y be in X. If $b \in B$, then

$$d(x, B) \le d(x, b) \le d(x, y) + d(y, b).$$

Passing to the infimum over B we obtain

$$d(x, B) \le d(x, y) + d(y, B).$$

Inverting the roles of x and y yields

$$d(y, B) \le d(y, x) + d(x, B) = d(x, y) + d(x, B)$$

and hence

$$|d(x, B) - d(y, B)| \le d(x, y),$$

from which we easily establish the uniform continuity of g.

SOLUTION 2.3.23.

(1) Recall that f is uniformly continuous on $\mathbb{R}^+$ iff

$$\forall \varepsilon > 0, \exists \alpha > 0, \ \forall x, y \in \mathbb{R}^+ \ (|x - y| < \alpha \Rightarrow d(f(x), f(y)) < \varepsilon).$$

Let $\varepsilon > 0$. Since f is the identity mapping, we can write

$$d(f(x), f(y)) = d(x, y) = |\sqrt{x} - \sqrt{y}|.$$

It is well-known that

$$\forall x, y \ge 0 : \ |\sqrt{x} - \sqrt{y}| \le \sqrt{|x - y|}.$$

So it becomes clear that in order to establish the uniform continuity of f, it suffices to take $\alpha = \varepsilon^2$.

(2) We could say that the function $g : \mathbb{R}^+ \to \mathbb{R}^+$ (both spaces equipped with the usual metric), defined for all $x \ge 0$, by $f(x) = \sqrt{x}$ is uniformly continuous on $\mathbb{R}^+$.

SOLUTION 2.3.24. The needed tools to answer these questions are the density of both $\mathbb{Q}$ and $\mathbb{R} \setminus \mathbb{Q}$ in $\mathbb{R}$.

(1) Consider

$$f(x) = \begin{cases} x - 1, & x \in \mathbb{Q}, \\ x + 1, & x \in \mathbb{R} - \mathbb{Q}. \end{cases}$$

Then f is not continuous at any point of $\mathbb{R}$. To show this, let $x_0 \in \mathbb{R}$. Then

$$\exists (x_n) \in \mathbb{Q}, (y_n) \in \mathbb{R} \setminus \mathbb{Q} : x_n \to x_0 \text{ and } y_n \to x_0.$$

If f were continuous at x_0, then would have

$$f(x_n) = x_n - 1 \to x_0 - 1 = x_0 + 1 \leftarrow y_n + 1 = f(y_n)$$

and obviously no such x_0 would satisfy that equation. Hence f is discontinuous on the whole of $\mathbb{R}$.

(2) Consider

$$f(x) = \begin{cases} 0, & x \in \mathbb{Q}, \\ x, & x \in \mathbb{R} - \mathbb{Q}. \end{cases}$$

Then f is only continuous at one point, that is at $x_0 = 0$. The proof is very similar to the previous one.

(3) Consider

$$f(x) = \begin{cases} x^2, & x \in \mathbb{Q}, \\ 2 - x, & x \in \mathbb{R} - \mathbb{Q}. \end{cases}$$

Then f is only continuous at two points, namely 1 and -2. The proof is also left to the reader for its resemblance to Answer 1.

SOLUTION 2.3.25. >From Exercise 2.3.22 we know that f is continuous. Now we have

$$B(a, r) = \{x \in X : d(x, a) < r\} = \{x \in X : f(x) < r\}$$
$$= \{x \in X : f(x) \in (-\infty, r)\}.$$

Hence

$$B(a, r) = f^{-1}((-\infty, r)).$$

Since $((-\infty, r)$ is open in $\mathbb{R}$ and f is continuous on X, then $B(a, r) = f^{-1}((-\infty, r))$ is open in X.

SOLUTION 2.3.26.

(1) The proof that δ is indeed a metric is left to the reader. The metric δ is bounded and hence $(\mathbb{R}, \delta)$ is bounded since

$$\forall x, y \in \mathbb{R} : \delta(x, y) = |\arctan x - \arctan y| < \pi.$$

(2) The range of the "arctan" function is $(-\frac{\pi}{2}, \frac{\pi}{2})$. This observation leads to

$$B(0,2) = B(0,4) = B(1,4) = \mathbb{R}.$$

Thus, two balls with different radii (and the same center) may coincide with respect to this metric. Also, two balls with the same radius (and different centers) may also be equal!

(3) The two metrics cannot be equivalent. Assume they were, then we would have for some $\beta > 0$ and all $x, y \in \mathbb{R}$:

$$\beta|x - y| \leq |\arctan x - \arctan y| < \pi.$$

This would imply then that (usual) $\mathbb{R}$ is bounded! This is impossible.

(4) Yes. This can be easily seen from the fact that the identities

$$\text{id} : (\mathbb{R}, \delta) \to (\mathbb{R}, d) \text{ and id} : (\mathbb{R}, d) \to (\mathbb{R}, \delta)$$

are both continuous for the functions arctan and tan are both continuous in usual $\mathbb{R}$.

SOLUTION 2.3.27.

(1) It is clear that for all $x, y \in X$

$$\rho(x, y) \leq d(x, y).$$

On the other hand, the ratio $\frac{\rho}{d}$ is not a bounded function on X^2, i.e. there is no positive constant a such that for all x, y $\rho(x, y) \geq d(x, y)$. Therefore, d and ρ are not equivalent metrics.

(2) Let us show that d and ρ are topologically equivalent. Denote an open ball in (X, d) by B_d and an open ball in (X, ρ) by B_ρ. Let U be an open set in (X, d). Let $x \in U$. Then $\exists r > 0$ such that $B_d(x, r) \subset U$. But

$$\frac{d(x, y)}{1 + d(x, y)} < \frac{r}{1 + r} \Rightarrow rd(x, y) + d(x, y) < rd(x, y) + r \Rightarrow d(x, y) < r$$

(for each $x, y \in X$ and $r > 0$). Hence

$$B_\rho\left(x, \frac{r}{1+r}\right) \subset B_d(x, r) \subset U,$$

proving the openness of U in (X, ρ).

Conversely, let U be an open set in (X, ρ). Let $x \in U$. Then $\exists r > 0$ such that $B_\rho(x, r) \subset U$. But, if $r < 1$, then

$$d(x, y) < \frac{r}{1 - r} \Rightarrow \frac{d(x, y)}{1 + d(x, y)} = \rho(x, y) < r$$

(for all x, y). Hence

$$B_d\left(x, \frac{r}{1-r}\right) \subset B_\rho(x, r) \subset (f^{-1})^{-1}(U) = U$$

and if $r \geq 1$, then for all $x, y \in X$

$$\frac{d(x, y)}{1 + d(x, y)} \leq 1 \leq r.$$

Hence $B_\rho(x, r) = X$ which forces us to have $X = U$. Thus, in either case U is open in (X, d). The proof is complete.

2.4. Hints/Answers to Tests

SOLUTION 1. Let $x, y \in X$. We have

$$0 = d(x, x) \leq d(x, y) + d(y, x) = d(x, y) + d(x, y) = 2d(x, y)$$

where we have used the three properties of a metric...

SOLUTION 2. Straightforward calculations based on known properties of the Logarithm function...

SOLUTION 3. Proceed as in Exercise 2.3.12. Also use the Cauchy-Schwarz inequality...

SOLUTION 4. It is obvious that d is positive. The first two properties are evidently verified. For the triangle inequality, there is a finite number of cases that must be checked...

SOLUTION 5. The second and the third properties use the fact that d is a metric but they do not use the injectivity of f...To prove the first property, utilize the injectivity of f...

SOLUTION 6. The discrete metric space is a ultrametric space whereas the usual metric on $\mathbb{R}$ is not...

CHAPTER 3

Topological Spaces

3.2. True or False: Answers

ANSWERS.

(1) The answer is yes. We may define at least two topologies on X, namely the discrete and the indiscrete ones.

Observe that if $\operatorname{card} X = 1$, then these topologies manifestly coincide.

(2) If we come to show that the intersection of two elements of T is in T, then a proof by induction will allow us to deduce that the finite intersection of elements in T is in T too. This is probably known to most of the readers, but we wanted to remind the students that the proof by induction is used here for finite intersections and it cannot be used for infinite countable unions or intersections.

(3) False! For example, define on $X = \{a, b, c, d\}$

$$T = \{\varnothing, \{a\}, X\} \text{ and } T' = \{\varnothing, \{b, d\}, X\}.$$

Then it can easily be established that T and T' are two topologies on X. However,

$$T \cup T' = \{\varnothing, \{a\}, \{b, d\}, X\}$$

does not define a topology on X.

(4) True! In fact, a more general result holds. Namely, if $\{T_i\}_{i \in I}$ is an *arbitrary* collection of topologies on the *same* set X, then so is their intersection. Let us prove it. Since T_i are all topologies for $i \in I$, they must all contain $\varnothing$ and hence so must do their intersection. The same reasoning applies for X. Let $A, B \in \bigcap_{i \in I} T_i$. Then

$$\forall i \in I : \ A, B \in T_i \text{ and thus for all } i, \ A \cap B \in T_i.$$

Therefore, $A \cap B \in \bigcap_{i \in I} T_i$. Finally, let $A_j \in \bigcap_{i \in I} T_i$ where $j \in J$.
Then all A_j belongs to each T_i. Since T_i are all topologies, it follows that $\bigcup_{j \in J} A_j \in T_i$, for all $i \in I$. The proof is complete.

(5) The answer is yes and both mappings are "increasing" with respect to "$\subset$". For the known properties state that

$$A \subset B \Longrightarrow \begin{cases} \overline{A} \subset \overline{B}, \\ \overset{\circ}{A} \subset \overset{\circ}{B}. \end{cases}$$

(6) The answer is yes if and only if $A = X$ (as X is open). And the answer is no and will always be so if $A \neq X$, i.e. $A \subsetneq X$ as $\overset{\circ}{A} \subset A \subsetneq X$.

(7) The answer is no! Give $\mathbb{R}$ the usual topology and take $A = \mathbb{Q}$. Then

$$\overset{\circ}{\overline{A}} = \overset{\circ}{\mathbb{R}} = \mathbb{R} \neq \varnothing.$$

Remember that a set A such that $\overset{\circ}{\overline{A}} = \varnothing$ is said to be nowhere dense.

(8) The answer is yes if and only if $A = \varnothing$ (as $\varnothing$ is closed). And the answer is no and will ever be so if $A \neq \varnothing$ for the simple reason that by definition of the closure of a set, we have $\overline{A} \supset A$.

(9) Absolutely not! The two terminologies are different from one another and are totally independent of each other.

(10) The answer is no. We give two counterexamples. Define a *topology* T on $X = \{a, b, c\}$ by $T = \{\varnothing, \{b\}, X\}$. Then X is evidently not Hausdorff in T. Now take $A = \{b\}$ which is Hausdorff while $\overline{\{b\}} = X$ is not. Another example (very similar in core though) is to take $A = \{a\}$ and the topology of Exercise 3.3.25.

(11) True. To see this, assume that X is a Hausdorff space and let $A \subset X$. Now, let $x, y \in A$ with $x \neq y$. By the Hausdorffness of X, there are two open sets U and V containing x and y respectively such that $U \cap V = \varnothing$. The proof is now complete as $x \in A \cap U$, $y \in A \cap V$ and

$$(A \cap U) \cap (A \cap V) = A \cap (U \cap V) = A \cap \varnothing = \varnothing.$$

(12) The answer is yes. We provide a proof. Let $x, y \in X$ be two distinct points in X. Since T is Hausdorff,

$$\exists U \in \mathcal{V}(x), \exists V \in \mathcal{V}(x) : U \cap V = \varnothing.$$

But, any open set in T is open in T' and hence T' is Hausdorff too.

(13) The answer is no! See Exercise 3.3.25.

(14) Yes and the indiscrete topological space is a good example. For if A is a nonvoid subset of a set X endowed with the indiscrete topology, then the smallest closed set containing A is X, i.e. $\overline{A} = X$.

(15) The answer is no. Endow $X = \mathbb{R}$ with the usual topology. Let $Y = (0, 2]$ be a topological subspace of X. Let $A = (1, 2]$. Then

$$\overset{\circ Y}{A} = (1, 2] \text{ and } \overset{\circ X}{A} = (1, 2)$$

and so

$$\overset{\circ Y}{A} = (1, 2] \neq Y \cap \overset{\circ X}{A} = (1, 2).$$

(16) The answer is no. Apart from some obvious sets such as a finite set say, it is not clear how one can introduce such a definition in an arbitrary topological space. In metric spaces, this is possible thanks to balls which, even if they are subsets of an arbitrary set, their definition depends on a positive number and hence one can impose some constraint on a set to be bounded. If the topological space is also given a structure of a vector space (more known as *topological vector spaces*), then one can introduce a definition of a bounded set. This will not be discussed in this book.

(17) Firstly, this is a purely algebraic question but it is of interest especially in the product topology.

   The answer is no! There are many counterexamples. Here is one: in $\mathbb{R} \times \mathbb{R}$, take

$$A = [0, \infty) \text{ and } B = [0, \infty)$$

Then

$$(A \times B)^c \neq A^c \times B^c! \text{ check it out!}$$

What is *true* is the following

$$(A \times B)^c = \{(x, y) \in X \times Y : x \notin A \text{ or } y \notin Y\} = (A^c \times Y) \cup (X \times B^c).$$

(18) The answer is no! Take $X = Y = \mathbb{R}$ both endowed with the usual topology. Then $U = [-1, 1)$ is neither closed nor open in X and yet $U \times \varnothing = \varnothing$ which is closed and open in $X \times Y$.

(19) The answer is yes! In euclidian $\mathbb{R}^2$, let $B((0, 0), 1)$ be the open ball (hence open) with center $(0, 0)$ and radius 1. Assume that

this open ball could be written as

$$B((0,0),1) = U \times V.$$

Then $(0.8,0) \in B((0,0),1)$ giving $0.8 \in U$ and $(0,0.7) \in B((0,0),1)$ giving $0.7 \in V$. However, $(0.8,0.7) \notin B((0,0),1)$.

There are many more examples such as: $\{(x,y) \in \mathbb{R}^2 : xy > 2\}$ or $\{(x,y) \in \mathbb{R}^2 : x + y < 1\}$.

(20) The answer is no! First we recall that the exterior of A is the interior of the complement of A. As a counterexample, take $A = \mathbb{Q}$ with respect to the usual topology of $\mathbb{R}$. Then

$$\overset{\circ}{\mathbb{Q}} \cup \overbrace{\mathbb{R} \setminus \mathbb{Q}}^{\circ} = \varnothing \cup \varnothing = \varnothing \neq X.$$

(21) The answer is no! In usual $X = \mathbb{R}$, take $A = \mathbb{Q}$. Then

$$\overset{\circ}{\mathbb{Q}} = \varnothing \neq \overset{\circ}{\overline{\mathbb{Q}}} = \overset{\circ}{\mathbb{R}} = \mathbb{R}.$$

(22) The answer is no! In usual $X = \mathbb{R}$, take $A = \mathbb{Q}$. Then

$$\overline{\mathbb{Q}} = \mathbb{R} \neq \overline{\overset{\circ}{\mathbb{Q}}} = \overline{\varnothing} = \varnothing.$$

(23) The answer is no! Let $A = \mathbb{N}$ in the usual topology of $\mathbb{R}$. Then A is closed (why?) but $A' = \varnothing$. We show this last result. Let x be any real number. Remember that

$$x \in \mathbb{N}' \iff \forall \varepsilon > 0, (x - \varepsilon, x + \varepsilon) \cap \mathbb{N} - \{x\} \neq \varnothing.$$

A quick observation tells us that the intersection intervening in the previous statement need not be non-empty for all ε.

The analogous question for open sets is true (see Exercise 3.3.8).

(24) False! Consider a non empty set X with the indiscrete topology. Let $A = \{a\}$ where $a \in X$. The only non-empty set is X and hence

$$x \in A' \iff X \cap A - \{x\} \neq \varnothing.$$

We clearly see that all points but a verifies the last equivalence and hence $A' = \{a\}^c$ which is not closed.

REMARK. In a metric space, every derived set is closed.

(25) The answer is yes. We provide simple examples with details to be seen below in some exercises. The topology here is the usual one.

(a) Let $A = \{\frac{1}{n} : n \geq 1\}$. Then $A' = \{0\}$ and hence $A \cap A' = \varnothing$.

(b) For instance, take $A = \{0\}$. Then $A' = \varnothing$. Thus $A' \subsetneq A$.

(c) If $A = (0,1)$, then $A' = [0,1]$ and hence $A \subset A'$.

(d) If $A = \mathbb{R}$, then $A' = \mathbb{R}$ and so $A = A'$.

(26) Only the implication "$\Rightarrow$" holds. Let us show that. Let $x \in A'$. This means that

$$\forall U \in \mathcal{V}(x) : \ U \cap A - \{x\} \neq \varnothing.$$

Since $A \subset B$, we immediately get that $A - \{x\} \subset B - \{x\}$. Hence

$$\varnothing \neq U \cap A - \{x\} \subset U \cap B - \{x\}$$

yielding $U \cap B - \{x\} \neq \varnothing$ or $x \in B'$.

The other implication is false in general. One possible example is to take $A = (0,1)$ and $B = [0,1]$. Then

$$A \subset B \text{ but } A' = B'.$$

(27) The answer is no! In usual $\mathbb{R}$, let $A = \mathbb{Q}$. Then

$$\mathrm{Fr}(\overline{\mathbb{Q}}) = \mathrm{Fr}(\mathbb{R}) = \mathbb{R} \setminus \mathring{\mathbb{R}} = \varnothing \neq \mathrm{Fr}(\mathbb{Q}) = \overline{\mathbb{Q}} \setminus \mathring{\mathbb{Q}} = \mathbb{R} \setminus \varnothing = \mathbb{R}.$$

(28) False! In usual $\mathbb{R}$, take $A = \mathbb{Q}$. Then

$$\mathrm{Fr}(\mathring{\mathbb{Q}}) = \varnothing \neq \mathrm{Fr}(\mathbb{Q}) = \mathbb{R}.$$

What always holds is: $\mathrm{Fr}(\mathring{A}) \subset \mathrm{Fr}(A)$. The proof is a mere consequence of $\mathring{\overline{A}} \subset \overline{A}$.

(29) False! In usual $\mathbb{R}$, take $A = [0,1]$ and $B = [1,2]$. Then

$$\mathrm{Fr}(A \cup B) = \mathrm{Fr}([0,2]) = \{0,2\} \neq \mathrm{Fr}([0,1]) \cup \mathrm{Fr}([1,2]) = \{0,1,2\}.$$

Even for disjoint A and B, the assertion has a negative answer. However, if $\overline{A}$ and $\overline{B}$ are disjoint, then $\mathrm{Fr}(A \cup B) = \mathrm{Fr}(A) \cup \mathrm{Fr}(B)$.

What always holds is: $\mathrm{Fr}(A \cup B) \subset \mathrm{Fr}(A) \cup \mathrm{Fr}(B)$. Here is a simple proof

$$\mathrm{Fr}(A \cup B) = (\overline{A \cup B}) \setminus (\mathring{\overline{A \cup B}}) = (\overline{A} \cup \overline{B}) \cap (\mathring{\overline{A \cup B}})^c$$

$$= (\overline{A} \cup \overline{B}) \cap (\mathring{\overline{A \cup B}})^c$$

$$\subset (\overline{A} \cup \overline{B}) \cap (\mathring{A} \cup \mathring{B})^c$$

$$\subset \mathrm{Fr}(A) \cup \mathrm{Fr}(B).$$

(30) The answer is again no! For in the usual topology of $\mathbb{R}$, $\mathbb{Q}$ verifies

$$\mathbb{Q} \subset \operatorname{Fr} \mathbb{Q} = \mathbb{R}.$$

(31) The answer is yes and this a simple consequence of the definition of the frontier, namely

$$\operatorname{Fr}(A) = \overline{A} \setminus \overset{\circ}{A} = \overline{A} \cap (\overset{\circ}{A})^c \subset \overline{A}.$$

(32) The answer is no! As the last but one answer, take $A = \mathbb{Q}$ and $B = \mathbb{R}$. Then $A \subset B$ whilst

$$\operatorname{Fr} A = \mathbb{R} \not\subset \operatorname{Fr} B = \varnothing.$$

REMARK. The hypothesis $A \subset B$ does not imply $\operatorname{Fr} B \subset \operatorname{Fr} A$ either. In the usual topology of $\mathbb{R}$, take $A = \mathbb{Q}^+$ and $B = \mathbb{Q}$. Then

$$A \subset B \text{ but } \operatorname{Fr} A = \mathbb{R}^+ \not\supset \operatorname{Fr} B = \mathbb{R}.$$

(33) None of the assertions holds! Let X be a non-empty set endowed with the discrete metric. Let $r = 1$ and let $x \in X$. Then we know from Exercise 2.3.5 that $S(x, 1) = X \setminus \{x\}$. Now since every set is simultaneously open and closed in a discrete metric space, we have

$$\operatorname{Fr}(B(x, 1)) = \varnothing \text{ and } \operatorname{Fr}(B_c(x, 1)) = \varnothing$$

and thus

$$\operatorname{Fr}(B(x, r)) \neq S(x, r) \text{ and } \operatorname{Fr}(B_c(x, r)) \neq S(x, r).$$

(34) A common definition of a neighborhood is: U is said to be a neighborhood of a ($a \in X$ and X is a topological space) if there is an open set V containing a such that $V \subset U$. If X is a metric space, then we may change V by an open ball centered at a.

According to this definition, $(-1, 1]$ is a neighborhood of 0 since $0 \in (-1, 1) \subset (-1, 1]$ and $(-1, 1)$ is an open set in $\mathbb{R}$. $(-2, 2)$ is also a neighborhood of 0. The other sets are not for:
(a) $(0, 2]$ does not even contain 0;
(b) $[-1, 0]$ cannot contain an open set which contain 0;
(c) $[0, 1]$ cannot contain an open set which contain 0.
But, according to the definition of this book (and considered by others such as [10] or [11]), only $(-2, 2)$ is a neighborhood of 0.

(35) First, $T \subset T'$ is usually referred to T' being finer than T or T being coarser than T'. In other words, the "finer" is the "fatter". Sometimes, the terms "stronger" and "weaker" instead of "finer" and "coarser" respectively and so are the terms "larger" and "smaller". The last two couples of terms are probably more meaningful than the first couple of terms. In any case, the reader should check the terminology used in each book. As for ours, we use the "finer-coarser" terminology.

Going back to the question. The first statement is true and it is in fact the definition of T' being finer than T. The second statement is false. Saying that T' is finer that T can also be interpreted using closed sets and in the same way, that is, if every closed set in T is closed in T', then we also say that T' is finer than T. The reason is simple, if V is closed in T, then V^c is open in T. Now if V^c is open in T' too, then V is closed in T'.

(36) False! For a counterexample see Exercise 3.3.38.

3.3. Solutions to Exercises

Solution 3.3.1.

- Possible topologies on X: The only possible topology here is $T = \{\varnothing, \{1\}\}$.
- Possible Topologies on Y: There are four possible topologies in this case. They are

$$T_1 = \{\varnothing, \{1\}\}; \ T_2 = \{\varnothing, \{1\}, Y\}; \ T_2 = \{\varnothing, \{2\}, Y\} \text{ and } T_4 = \mathcal{P}(Y).$$

Solution 3.3.2.

(1) Obviously X and $\varnothing$ are both in T and the reader may easily check that finite intersections of elements of T is again in T and that the arbitrary union (finite in this exercise) of elements of T is in T too.

(2) The closed sets in this case are easy. They are the complements of the open sets, i.e. the complements of the elements of T. Hence they are

$$X, \ \{b, c, d, e\}, \ \{a, b, e\}, \ \{b, e\}, \ \{a\}, \ \varnothing.$$

(3) The closure of $\{a\}$ is the smallest closed set containing $\{a\}$ (which is also the intersection of all closed sets containing $\{a\}$). Hence $\overline{\{a\}} = \{a\}$ (or here simply since $\{a\}$ is closed!). In a similar way we find that $\overline{\{b\}} = \{b, e\}$.

The interior of $\{a\}$ is the biggest open set contained in $\{a\}$.

Hence $\overset{\circ}{\widehat{\{a\}}} = \{a\}$ (or just because $\{a\}$ is open). Similarly $\overset{\circ}{\widehat{\{b\}}} = \varnothing$. Hence

$$\text{Fr}(\{a\}) = \overline{\{a\}} \setminus \overset{\circ}{\widehat{\{a\}}} = \varnothing$$

and

$$\text{Fr}(\{b\}) = \overline{\{b\}} \setminus \overset{\circ}{\widehat{\{b\}}} = \{b, e\}.$$

(4) The smallest closed set containing $\{a, b\}$ is X. Thus $\overline{\{a, b\}} = X$, i.e. $\{a, b\}$ is dense in X.

(5) A neighborhood of c is any open set containing c (remember that this is the definition adopted in this book). Hence

$$\mathcal{V}(c) = \{\{c, d\}, \{a, c, d\}, \{b, c, d, e\}, X\}.$$

Also,

$$\mathcal{V}(d) = \{\{c, d\}, \{a, c, d\}, \{b, c, d, e\}, X\}.$$

(6) No T is not Hausdorff since

$$\exists c, d \in X \ (c \neq d), \forall (U, V) \in \mathcal{V}(c) \times \mathcal{V}(d) : U \cap V \neq \varnothing.$$

SOLUTION 3.3.3.

(1) $\varnothing \in T$ (by definition) while $\mathbb{N} \in T$ for $n = 1$. The other axioms of a topological are the matter of easy unions and intersections and hence left to the reader (for the arbitrary union one has just to justify the existence of $\inf\{n_p : p \in \mathbb{N}\}$).

(2) No, the given set is not open (why?).

(3) The only two open sets containing 2 are $\{1, 2, 3, \cdots\}$ and $\{2, 3, 4, \cdots\}$. Hence

$$\mathcal{V}(2) = \{\{1, 2, 3, \cdots\}, \{2, 3, 4, \cdots\}\}.$$

Similarly

$$\mathcal{V}(3) = \{\{1, 2, 3, \cdots\}, \{2, 3, 4, \cdots\}, \{3, 4, 5 \cdots\}\}.$$

(4) No, T is not Hausdorff. The previous question provides us with a counterexample (doesn't it?).

(5) The closed sets are of the form $\{1, 2, 3, \cdots, p\}$ where $p \in \mathbb{N}$.

(6) The interior of $\{4\}$ is the biggest open set contained in it, that is the empty set! The smallest closed set containing $\{4\}$, i.e. its closure, is $\{1, 2, 3, 4\}$.

The interior and the closure of $\{2, 4, 6, 8, \cdots\}$ are respectively $\varnothing$ and $\mathbb{N}$.

(7) We claim that a set A is dense in $\mathbb{N}$ iff it is infinite. We first note that if A is finite, then it cannot be dense as it is closed in this case and so we have proved the implication "$\Rightarrow$". Now, if A is infinite, then the smallest closed superset is $\mathbb{N}$ and thus we have proved the implication "$\Rightarrow$".

SOLUTION 3.3.4. In order to show that T is the discrete topology, it suffices to check that every subset of X is open in T. Let A be a subset of X. Then A can always be written as a union (finite in this case) of those singletons. We have

$$\{a, b\} = \{a\} \cup \{b\}, \{a, c\} = \{a\} \cup \{c\}, \ldots\ldots, \{a, b, c\} = \{a\} \cup \{b\} \cup \{c\}.$$

Thus, any subset is open (and closed).

REMARK. The previous proof applies to show that any topological space in which singletons are open (hence/or closed) is a discrete space.

SOLUTION 3.3.5. Recall that by definition, the interior of a set A is the largest open set contained in A (which may be taken to be the union of all open sets contained in A).

Similarly, the closure of a set A is the smallest closed set containing A (which may be taken to be the intersection of all closed sets containing A).

We note that if U is an open set *contained* in A, then $U^c := V$ is a closed set *containing* A^c. These remarks allow us to easily give the solution to this exercise and we have for open sets U

$$(\overset{\circ}{A})^c = \left(\bigcup_{U \subset A} U \right)^c = \bigcap_{U \subset A} U^c = \bigcap_{A^c \subset V} V = \overline{A^c}.$$

The other property can be easily established by replacing A by A^c.

SOLUTION 3.3.6. Well, the right-to-left implication is trivial (is it not?). To show the other implication, assume that $\overline{A} \cap U \neq \varnothing$. Then there is some x in both U and $\overline{A}$. Hence for any open set V containing x, $V \cap A \neq \varnothing$. In particular, taking $V = U$ (which is legitimate as U is open and it contains x) gives $A \cap U \neq \varnothing$.

Another (simpler) proof goes as follows. If $A \cap U = \varnothing$, then $A \subset U^c$. But U^c is closed and hence $\overline{A} \subset U^c$ or $\overline{A} \cap U = \varnothing$.

SOLUTION 3.3.7.

(1) Let us find the closure of $\mathbb{Q}$ in $\mathbb{R}$. We know that $\mathbb{Q} \subset \mathbb{R}$ and hence $\overline{\mathbb{Q}} \subset \overline{\mathbb{R}} = \mathbb{R}$. Let us show that $\mathbb{R} \subset \overline{\mathbb{Q}}$. Let $x \in \mathbb{R}$. By

definition

$$x \in \overline{\mathbb{Q}} \iff \forall \varepsilon > 0, \ (x - \varepsilon, x + \varepsilon) \cap \mathbb{Q} \neq \varnothing.$$

But, any interval contains always rational numbers and hence $\forall \varepsilon > 0, \ (x - \varepsilon, x + \varepsilon) \cap \mathbb{Q} \neq \varnothing$. Hence $\mathbb{R} \subset \overline{\mathbb{Q}}$ and thus $\overline{\mathbb{Q}} = \mathbb{R}$. In a very similar way (with an obvious change) one can show that $\overline{\mathbb{R} \setminus \mathbb{Q}} = \mathbb{R}$.

We now show that $\overline{(0, 1]} = [0, 1]$. To see this, we observe that 0 belongs to $\overline{(0, 1]}$ as

$$\forall \varepsilon > 0, \ (-\varepsilon, \varepsilon) \cap (0, 1] \neq \varnothing.$$

Any other point $x \notin [0, 1]$, i.e. $x < 0$ or $x > 1$, does not lie in $\overline{(0, 1]}$ since we can easily find a small enough $\varepsilon > 0$ such that $(x - \varepsilon, x + \varepsilon) \cap (0, 1] = \varnothing$. Thus $\overline{(0, 1]} = [0, 1]$.

We can show as before that $\overline{(2, 3]} = [2, 3]$. Besides since $\{1\}$ is closed in $\mathbb{R}$ (because its complement in $\mathbb{R}$, being $(-\infty, 1) \cup (1, +\infty)$, is open since it is a union of open sets), one has $\overline{\{1\}} = \{1\}$. Hence

$$\overline{\{1\} \cup (2, 3]} = \overline{\{1\}} \cup \overline{(2, 3]} = \{1\} \cup [2, 3]$$

(2) The interiors of $\mathbb{Q}$ and $\mathbb{R} \setminus \mathbb{Q}$ are both empty for a similar reason. We know that

$$x \in \overset{\circ}{\mathbb{Q}} \iff \exists r > 0 : \ (x - r, x + r) \subset \mathbb{Q}.$$

But, since an interval contains irrational numbers,

$$\forall r > 0 : \ (x - r, x + r) \not\subset \mathbb{Q}.$$

Thus $\overset{\circ}{\mathbb{Q}} = \varnothing$. Also, since an interval contains rational numbers, we obtain

$$\overset{\circ}{\overbrace{\mathbb{R} \setminus \mathbb{Q}}} = \varnothing.$$

Let us find now $\overset{\circ}{\overbrace{(0, 1]}}$. First, we note that $1 \notin \overset{\circ}{\overbrace{(0, 1]}}$ since if it were, we would have for some $r > 0$, $(1 - r, 1 + r) \in (0, 1]$ which is obviously not the case for any $r > 0$. Now, any x in $(0, 1)$ is interior to $(0, 1]$. Let $x \in (0, 1)$, then we can always find $0 < r < \min(x, 1 - x)$ such that $(1 - r, 1 + r) \in (0, 1]$.

Finally, if $x \notin (0, 1]$, then $x \notin \overset{\circ}{\overbrace{(0, 1]}}$. Therefore,

$$\overset{\circ}{\overbrace{(0, 1]}} = (0, 1).$$

(3) The answer is no for both cases. We give counterexamples.
Consider $A_n = \left(-\frac{1}{n}, \frac{1}{n}\right)$, $n \geq 1$. Then $\overset{\circ}{A_n} = \left(-\frac{1}{n}, \frac{1}{n}\right)$. Hence

$$\left(\bigcap_{n \geq 1} A_n\right)^{\circ} = \overset{\circ}{\overbrace{\{0\}}} = \varnothing \neq \bigcap_{n \geq 1} \overset{\circ}{A_n} = \{0\}.$$

For the other equality, consider $A_n = \left[\frac{1}{n}, 1\right]$, $n \geq 1$. Then

$$\overline{\bigcup_{n \geq 1} A_n} = \overline{(0, 1]} = [0, 1] \neq \bigcup_{n \geq 1} \overline{A_n} = (0, 1].$$

(4) Yes $\mathbb{R}$ is separated. To see this take any two real numbers x and y such that $x \neq y$. We need to find two disjoint open neighborhoods of x and y. We may assume WLOG that $x > y$. Take the following open intervals

$$U = \left(\frac{x + y}{2}, +\infty\right) \text{ and } V = \left(-\infty, \frac{x + y}{2}\right).$$

Then $x \in U$ and $y \in V$ but $U \cap V = \varnothing$. Thus $\mathbb{R}$ is Hausdorff.

SOLUTION 3.3.8. Let a be a point in A. Since A is open,

$$\exists r > 0, \ B(a, r) \subset A.$$

We need to show that a is also a limit point of A, i.e. $a \in A'$, i.e.

$$\forall \varepsilon > 0, \ B(a, \varepsilon) \cap A \setminus \{a\} \neq \varnothing.$$

Let $\varepsilon > 0$. Choose a point b such that $d(a, b) = \frac{1}{3}\min(r, \varepsilon)$. Then $b \in B(a, \varepsilon)$ and also $b \in B(a, r) \subset A$. Since $a \neq b$, $b \in A \setminus \{a\}$. Thus $B(a, \varepsilon) \cap A \setminus \{a\}$ is non-empty. The solution is complete.

SOLUTION 3.3.9.

(1) First, $\overset{\circ}{A} = \varnothing$. One way of seeing this is the following

$$A \subset \mathbb{Q} \Longrightarrow \overset{\circ}{A} \subset \overset{\circ}{\mathbb{Q}} = \varnothing \Longrightarrow \overset{\circ}{A} = \varnothing.$$

All points of A are isolated. For instance 1 is an isolated point. For we can easily choose an $r > 0$ (for instance $r = \frac{1}{3}$) such that

$$(1 - r, 1 + r) \cap A \setminus \{1\} = \varnothing.$$

The only limit point is 0. By the Archimedes theorem, for all $\varepsilon > 0$ there exists $n \in \mathbb{N}$ such that $\frac{1}{n} < \varepsilon$. Hence

$$\forall \varepsilon > 0 : (-\varepsilon, \varepsilon) \cap A \setminus \{0\} \neq \varnothing.$$

(2) Since 0 is the only limit point, we have

$$\overline{A} = A \cup A' = \left\{ \frac{1}{n} : n \geq 1 \right\} \cup \{0\}.$$

Thus A is not closed and since $\overset{\circ}{A} = \varnothing$, $\mathrm{Fr}(A) = \overline{A}$.

(3) We need to verify that $\overset{\circ}{\overline{A}} = \varnothing$. If x were a point in $\overset{\circ}{\overline{A}}$, then there would exist an $r > 0$ such that $(x - r, x + r) \subset \overline{A}$ which obviously does not hold.

SOLUTION 3.3.10. In the usual topology of $\mathbb{R}$, the following set

$$A = (0, 1) \cup (1, 2] \cup \{3\}$$

will do as the reader can easily check that

$$\overset{\circ}{A} = (0, 1) \cup (1, 2),$$

$$\overline{A} = [0, 1] \cup [1, 2] \cup \{3\} = [0, 2] \cup \{3\},$$

$$\overline{\overset{\circ}{A}} = (0, 2),$$

and

$$\overset{\circ}{\overline{A}} = [0, 1] \cup [1, 2] = [0, 2],$$

i.e. the five sets are mutually different.

SOLUTION 3.3.11. We show that $A' = \varnothing$. Since X is discrete, $\{x\}$ is an open neighborhood of $x \in X$. So

$$\exists U = \{x\} \in \mathcal{V}(x), \ U \cap A - \{x\} = \varnothing,$$

i.e. $A' = \varnothing$.

SOLUTION 3.3.12.

(1) First, since A is a non-empty and bounded set of $\mathbb{R}$, both $\inf A$ and $\sup A$ exist. We know that

$$a = \inf A \Longleftrightarrow \begin{cases} \forall x \in A : \ x \geq a, \\ \forall \varepsilon > 0, \exists x_\varepsilon \in A : \ a \leq x_\varepsilon < a + \varepsilon. \end{cases}$$

Hence $x_\varepsilon \in [a, a + \varepsilon) \subset (a - \varepsilon, a + \varepsilon)$. We also know that in $\mathbb{R}$

$$a \in \overline{A} \Longleftrightarrow \forall \varepsilon > 0, \ (a - \varepsilon, a + \varepsilon) \cap A \neq \varnothing.$$

So let $\varepsilon > 0$. Then there exists $x_\varepsilon \in A$ such that $x_\varepsilon \in (a - \varepsilon, a + \varepsilon)$ and therefore $(a - \varepsilon, a + \varepsilon) \cap A \neq \varnothing$. Thus $a \in \overline{A}$. The proof for $\sup A$ can be dealt with similarly.

(2) The answer is no! Take $A = (0, 1)$ in the discrete topology of $\mathbb{R}$, say. Then A is closed, i.e. $\overline{A} = A$. However,

$$\inf A = 0 \notin \overline{A} = (0, 1) \text{ and } \sup A = 1 \notin \overline{A} = (0, 1).$$

(3) No! In usual $\mathbb{R}$, let $A = \{0\}$. Then $\overset{\circ}{A} = \varnothing$ and hence

$$\sup A = \inf A = 0 \notin \overset{\circ}{A}.$$

SOLUTION 3.3.13.

(1) First, we justify the existence of $\sup \overline{A}$. Since A is bounded above, for some m, M $A \subset (m, M]$ (M being real and m any number small than M and it may even be $-\infty$). Depending on m, we then have $\overline{A} \subset (m, M]$ or $\overline{A} \subset [m, M]$. So, in either case, $\overline{A}$ is bounded above. Since it is also non empty (why?), $\sup \overline{A}$ exists. Obviously, we have

$$A \subset \overline{A} \Longrightarrow \sup A \leq \sup \overline{A}.$$

Let us prove the other inequality. Let $M = \sup A$ and let $x > M$. Setting $r = \frac{x-M}{2}$ (then $r > 0$), we claim that

$$A \cap (x - r, x + r) = \varnothing.$$

To see this, assume there is some a in A such that $a \in (x - r, x + r)$. Hence $a > x - r > M$ and then a would bigger than $M = \sup A$, a clear contradiction. Thus $x \notin \overline{A}$ and hence M is an upper bound for $\overline{A}$. This certainly leads to

$$\sup \overline{A} \leq M = \sup A,$$

completing the proof.

(2) The main point is that the existence of $\sup A$ does not imply any more that of $\sup \overline{A}$ in another topological space (besides what does "bounded" mean in an arbitrary topological space?).

In $\mathbb{R}$ equipped with the indiscrete topology, take $A = [0, 1]$. Then $\sup A = 1$. Since A is dense in $\mathbb{R}$, $\sup \overline{A} = +\infty$ (or it does not exist as some prefer to say).

(3) First, and on the contrary to the "closure case", $\sup \overset{\circ}{A}$ may not even exist even if $\sup A$ exists and in the usual topology setting! For example, take $A = \{1\}$. Then $\overset{\circ}{A} = \varnothing$ and so

$$\sup \overset{\circ}{A} = -\infty \neq \sup A = 1.$$

Now assume $\sup \overset{\circ}{A}$ exists. Then it need not be equal to $\sup A$. In usual $\mathbb{R}$, let $A = (-1, 1) \cup \{2\}$. Then

$$\sup A = 2 \neq \sup \overset{\circ}{A} = \sup(-1, 1) = 1.$$

REMARK. We say a few words about $\sup \varnothing = -\infty$. It is known that if A is a bounded subset, then $\inf A \leq \sup A$ iff A is *non-empty*. So, since $\varnothing \subset \mathbb{R}$, every element of $\mathbb{R}$ is an upper bound for $\varnothing$ and the least upper bound is then $-\infty$. Similarly, every element of $\mathbb{R}$ is a lower bounded for $\varnothing$ and biggest among them is "$+\infty$", i.e. $\inf \varnothing = +\infty$. This, thankfully, agrees with what we recalled above, that is, A is empty iff $\sup A < \inf A$.

SOLUTION 3.3.14.

(1) To show the inclusion $\overline{B(x, r)} \subset B_c(x, r)$, we can show equivalently that $(B_c(x, r))^c \subset (\overline{B(x, r)})^c$. Let $y \in (B_c(x, r))^c$, i.e. $d(x, y) > r$. We need to find some open ball (containing y) which does not intersect $B(x, r)$. Set $s = d(x, y) - r > 0$. Hence $B(y, s) \cap B(x, r) = \varnothing$ (to show this take z in $B(y, s) \cap B(x, r)$ and find a contradiction). This means that $y \notin \overline{B(x, r)}$, i.e. $y \in (\overline{B(x, r)})^c$.

(2) Let $X = [-1, 0] \cup [1, 2]$ considered as a metric *subspace* (the associated metric being the standard one). We have

$$B(1, 1) = \{x \in X : |x - 1| < 1\} = ([-1, 0] \cup [1, 2]) \cap (0, 2) = [1, 2).$$

Hence $\overline{B(1, 1)} = \overline{[1, 2)} = [1, 2]$ while

$$B_c(1, 1) = \{x \in X : |x - 1| \leq 1\} = ([-1, 0] \cup [1, 2]) \cap [0, 2] = \{0\} \cup [1, 2].$$

Thus $B_c(1, 1) \not\subset \overline{B(1, 1)}$.

We give another example. Let X be a set with $\mathrm{card} X \geq 2$ (why?). Let us associate with X the discrete metric. Let $x \in X$. We know that $B(x, 1) = \{x\}$ and since every subset in a discrete metric space is closed (and open!), we get $\overline{B(x, 1)} = \overline{\{x\}} = \{x\}$. This on the one hand, and on the other hand

$$B_c(x, 1) = \{y \in X : d(x, y) \leq 1\} = X \not\subset \overline{B(x, 1)}.$$

REMARK. There are cases where $\overline{B(x, r)} = B_c(x, r)$ holds in metric spaces. For instance, this is true in $\mathbb{R}^2$ endowed with the euclidian metric (or just $\mathbb{R}$ with the standard metric).

However, in the setting of normed vector spaces (not considered in this book), we *always* have $\overline{B(x, r)} = B_c(x, r)$.

SOLUTION 3.3.15. Since $\varnothing, Y \in T$, $f^{-1}(\varnothing) = \varnothing$ and $f^{-1}(Y) = X$, we conclude that $\varnothing, X \in T'$.

Now, let V_i be in T' for all $i \in I$. Then for some U_i in T, $V_i = f^{-1}(U_i)$. Hence

$$\bigcup_{i \in I} V_i = \bigcup_{i \in I} f^{-1}(U_i) = f^{-1}\left(\bigcup_{i \in I} U_i\right)$$

belongs to T' because $\bigcup_{i \in I} U_i \in T$.

In the end, let V_1 and V_2 be in T'. So there are U_1 and U_2 in T such that $f^{-1}(U_1) = V_1$ and $f^{-1}(U_2) = V_2$. Whence

$$V_1 \cap V_2 = f^{-1}(U_1) \cap f^{-1}(U_2) = f^{-1}(U_1 \cap U_2) \in T'$$

as $U_1 \cap U_2 \in T$. The proof is complete.

SOLUTION 3.3.16. The proof is essentially very similar to the one just before. First, $\varnothing$ and A both belongs to T_A as

$$\varnothing = A \cap \varnothing \text{ and } A = A \cap X.$$

Second, let $\{V_i\}_{i \in I}$ be a collection in T_A. Then U_i may be written as $V_i = A \cap U_i$ for some V_i and hence

$$\bigcup_{i \in I} V_i = \bigcup_{i \in I} (A \cap U_i) = A \cap \left(\bigcup_{i \in I} U_i\right) \in T_A$$

since $\bigcup_{i \in I} U_i \in T$.

Finally, let V_1 and V_2 be in T_A. Then there are U_1 and U_2 such that $V_1 = A \cap U_1$ and $V_2 = A \cap U_2$. Therefore,

$$V_1 \cap V_2 = (A \cap U_1) \cap (A \cap U_2) = A \cap (U_1 \cap U_2)$$

is in T_A as $U_1 \cap U_2 \in T$.

SOLUTION 3.3.17.

(1) Yes the set $\{3\}$ is open in $A = [0, 1) \cup \{3\}$. To show this we need to write it as an intersection of A and an open set in $\mathbb{R}$. One possible choice is the following

$$\{3\} = ([0, 1) \cup \{3\}) \cap (2, 5).$$

(2) The answer is again yes as one can do the following

$$[0, 1) = [0, 1] \cap \underbrace{(-1, 1)}_{\text{open in } \mathbb{R}} \text{ and } (0, 1) = [0, 1] \cap \underbrace{(0, 1)}_{\text{open in } \mathbb{R}}.$$

(3) Yes indeed. Write

$$\{n\} = \mathbb{N} \cap (n - 1, n + 1).$$

(4) Both $[0,1]$ and $(2,3)$ are open since

$$[0,1] = A \cap (-1,2) \text{ and } (2,3) = A \cap (2,3).$$

We can deduce that $[0,1]$ and $(2,3)$ are also closed (in A!) too. This is simply because the complement of $[0,1]$ in A is $(2,3)$ and vice versa.

(5) The closure of $\left(0,\frac{1}{2}\right)$ in $\mathbb{R}$ is $\left[0,\frac{1}{2}\right]$. Hence the closure of $\left(0,\frac{1}{2}\right)$ in A is

$$\left(0,\frac{1}{2}\right)^A = \left[0,\frac{1}{2}\right] \cap A = \left(0,\frac{1}{2}\right].$$

(6) Let us denote the subspace topology on A by T_A. It is defined as

$$T_A = \{A \cap U : U \in T\}.$$

Hence we can find T_A explicitly and we have

$$T_A = \{\varnothing, \{c,d\}, A\}.$$

We observe that the smallest closed set containing $\{b,d\}$ is A. So $\overline{\{b,d\}}^A = A$ (i.e. $\{b,d\}$ is dense in A). We can also obtain the same result using the relativity of closures as follows

$$\overline{\{b,d\}}^A = \overline{\{b,d\}}^X \cap A = \{b,c,d,e\} \cap \{b,c,d\} = \{b,c,d\} = A.$$

SOLUTION 3.3.18.

(1) Yes X is clopen in A. Since $[\sqrt{2},\pi]$ is closed in $\mathbb{R}$, X is closed in A. Also, since $\sqrt{2}, \pi \notin \mathbb{Q}$, one can write

$$X = A \cap [\sqrt{2},\pi] = A \cap (\sqrt{2},\pi).$$

As $(\sqrt{2},\pi)$ is open in $\mathbb{R}$, then X is open in A, establishing the "clopenness" of X in A.

(2) Yes Y is clopen in B. Since $[0,2]$ is closed in $\mathbb{R}$, so is Y in B. Now since $0,2 \notin \mathbb{R} \setminus \mathbb{Q}$, we can write

$$Y = B \cap [0,2] = B \cap (0,2),$$

meaning that Y is open in Y as well.

(3) Yes Z is clopen in A. This is easily seen from

$$Z = A \cap [\sqrt{2},\pi) = A \cap (\sqrt{2},\pi) = A \cap [\sqrt{2},\pi].$$

(4) Z' is not clopen, more precisely, it is neither closed nor open. For example, if it were open we would have: for all $x \in Z'$, there is some $r > 0$ such that $B(x,r)$ (the open ball in B) is

contained in Z'. In particular, for $x = \sqrt{2}$ there corresponds an $r > 0$ such that

$$B(\sqrt{2}, r) = \{y \in B : |\sqrt{2} - y| < r\} = B \cap (\sqrt{2} - r, \sqrt{2} + r) \subset Z',$$

which does not hold as

$$\forall r > 0 : \ B \cap (\sqrt{2} - r, \sqrt{2} + r) \not\subset Z'.$$

SOLUTION 3.3.19. Since $[0,1] \cap \mathbb{Q} \subset [0,1]$, then $\overline{[0,1] \cap \mathbb{Q}} \subset \overline{[0,1]} = [0,1]$. Now let $x \in [0,1]$, then obviously

$$\forall \varepsilon > 0 : \ [0,1] \cap \mathbb{Q} \cap (x - \varepsilon, x + \varepsilon) \neq \varnothing.$$

Thus $x \in \overline{[0,1] \cap \mathbb{Q}}$.

SOLUTION 3.3.20.

(1) The elements of X are real numbers in $[0,1]$ whose digits are constituted of the numbers 3 and/or 5.
(2) The following observation

$$X \cap (0.3\overline{5}, 0.5\overline{3}) = \varnothing \ (\text{why?})$$

shows that X cannot be dense in $[0,1]$.

SOLUTION 3.3.21. We give three ways of answering this question.

(1) Since X is not Hausdorff, it cannot be metrizable.
(2) This is somehow similar to the previous method. Let $B(x, r)$ be the open ball of center x and radius $r = \frac{d(x,y)}{3} > 0$ ($x \neq y$). It is an open set in X. However, $B(x, r)$ contains x, i.e. $B(x, r) \neq \varnothing$ and $B(x, r)$ does not contain y, i.e. $B(x, r) \neq X$. Thus $B(x, r)$ is another open set in X. This clearly leads to a contradiction. Therefore, X is not metrizable.
(3) If there is a metric that induces the topology of X, then X and $\varnothing$ are not the only closed sets in X since every finite set is closed in a metric space. Accordingly, X cannot be metrizable.

SOLUTION 3.3.22.

(1) Let us show that T is in effect a topology on $\mathbb{R}$. First $\varnothing$ belongs to T by definition while $\mathbb{R}$ belongs to T since its complement is finite (it is the empty set!).

Let U and V be two elements of T, i.e. U^c and V^c are both finite (we assume both U and V are not empty, otherwise this is obvious). We need to show that $U \cap V$ belongs to T, i.e. $(U \cap V)^c$ is finite. But $(U \cap V)^c = U^c \cup V^c$ and it is finite. Hence $U \cap V$ does belong to T.

Now let $\{U_i\}_{i \in I}$ be an arbitrary collection of elements of T. If $U_i = \varnothing$ for all $i \in I$, then their union is the empty set and hence it belongs to T. If at least one of these elements is not empty, call it U_j (hence U_j^c is finite), then we can write

$$\left(\bigcup_{i \in I} U_i \right)^c = \bigcap_{i \in I} U_i^c \subset U_j^c$$

and hence $\left(\bigcup_{i \in I} U_i \right)^c$ is finite. This finishes the answer.

(2) Obviously $\varnothing$ and $\mathbb{R}$ are closed. The only other closed sets are the finite ones. For if A is finite, then it is closed as A^c is open because its complement (which is A) is finite. And if A is closed, then A^c is open and hence $(A^c)^c = A$ is finite.

(3) Every finite subset is closed in standard $\mathbb{R}$. There are many infinite subsets which are closed in the standard $\mathbb{R}$ (for example every interval of the type $[a, b]$). Thus the standard topology has more closed sets, so it is finer than T.

(4) It is not Hausdorff since if it were, then for any $x, y \in \mathbb{R}$ such that $x \neq y$ there would exist two open neighborhoods $U \in \mathcal{V}(x)$ and $V \in \mathcal{V}(y)$ (belonging to T) such that $U \cap V = \varnothing$. Hence $U^c \cup V^c = \varnothing^c = \mathbb{R}$ which contradicts the finiteness of both U^c and V^c.

(5) No, since a metrizable space has to be Hausdorff.

(6) (a) If A is finite, then it is closed (see Question 2) and hence $\overline{A} = A$.
As for $\overset{\circ}{A}$ we will show that it equals the empty set. Let B be an open set contained in A. Then A^c is contained in B^c and since B^c is finite, so is A^c and hence $\mathbb{R} = A \cup A^c$ would have to be finite! which is impossible. Hence $\overset{\circ}{A} = \varnothing$.

(b) If A is infinite, two cases must be looked at.

(i) If A^c is finite, then A is open and hence $\overset{\circ}{A} = A$. Also since A^c is finite, then from the previous question A^c has an empty interior and thus Exercise 3.3.5 yields

$$\left(\overline{A} \right)^c = \left(\overset{\circ}{A^c} \right) = \varnothing \Longrightarrow \overline{A} = \mathbb{R},$$

i.e. A is dense in $\mathbb{R}$ in this case.

(ii) Now the case A^c infinite. If $B \subset A$ where B is open, then $A^c \subset B^c$ which is impossible since B^c is finite, leading to $\overset{\circ}{A} = \varnothing$.

Now, since the only closed set which can contain the infinite set A is X, we deduce immediately that $\overline{A} = X$.

(7) Yes, $\mathbb{R}$ is separable since $\mathbb{Q}$ is a countable subset of $\mathbb{R}$ which satisfies the hypotheses of Question (5-b-i).

(8) If X is finite, then T becomes the discrete topology for a simple reason. That is, since X is finite, any subset A of X will have a finite complement. Hence every subset is open and thus every subset is closed too.

SOLUTION 3.3.23. No, T is not a topology on $\mathbb{R}$. Assume it is and take the two sets $U = (-\infty, 0)$ and $V = (0, \infty)$. They are both open in T since their complements are infinite. However,

$$(U \cup V)^c = U^c \cap V^c = \{0\} \text{ is finite and hence } U \cup V \notin T.$$

Thus S is a not a topological space.

SOLUTION 3.3.24.

(1) Yes T is indeed a topology on X. For a change and also to facilitate the proof we will prove that T is a topology using closed sets and hence we need to show that $\varnothing$ and X are elements of T (which is obvious here), that the finite union of closed sets and the arbitrary intersection of closed sets are all in T. If we want a set to be closed in T, then it has to be countable. We are done as we know that the *finite* union and the *arbitrary* intersection of countable sets remain countable.

(2) No. If it were, X which is uncountable, would be a union of two countable sets!! (cf. Exercise 3.3.22).

(3) Remember that the closed sets are the countable ones (together with X and $\varnothing$). Now if A is a closed set different from X, then it is countable and so will any subset of A be.

(4) Let $\{U_n\}_n$ be a countable family of open sets in T. To show $\bigcap_{n=1}^{\infty} U_n$ is open, i.e. $\left(\bigcap_{n=1}^{\infty} U_n \right)^c$ is closed, i.e. countable. But

$$\left(\bigcap_{n=1}^{\infty} U_n \right)^c = \bigcup_{n=1}^{\infty} U_n^c$$

which is countable as a countable union of countable sets.

This result may fail to hold in usual $\mathbb{R}$ as shown by the classical example $U_n = (-\frac{1}{n}, \frac{1}{n})$, $n \in \mathbb{N}$.

(5) The proof is very similar to that of the non-Hausdorffness of the space. Assume $U_1 \cap U_2 \cap \cdots \cap U_n = \varnothing$ where the U_i ($i = 1, 2, \cdots, n$) are all open. Then $U_1^c \cup U_2^c \cup \cdots \cup U_n^c = \mathbb{R}$ which is impossible since the left hand side is countable while the right hand side is not. Thus the finite intersection of open sets is non-empty.

This result is not true in usual $\mathbb{R}$ in general. Consider for instance $(0, 1)$ and $(2, 3)$.

(6) No! Since $\mathbb{Q}$ is countable, it is closed. Thus $\overline{\mathbb{Q}} = \mathbb{Q} \neq \mathbb{R}$.

We claim that every uncountable set is dense in $\mathbb{R}$: To see this, let A be an uncountable subset of $\mathbb{R}$. Then A is not closed. Besides, A cannot be a subset of any closed set apart from X (why?). Thus A is dense in $\mathbb{R}$ equipped with this topology and hence so are the two given sets.

(7) Any *countable* set will make the topology T discrete. For if X is countable, $\{x\}$ ($x \in X$) will be clopen as $\{x\}$ and $X - \{x\}$ are both closed since they are both countable.

SOLUTION 3.3.25.

(1) It is a routine by now and it is left to the interested reader.

(2) No T is not Hausdorff. Let $x, y \in X$ such that $x \neq y$ (x or y may be worth a). Let U and V be two open sets containing x and y respectively. Then they contain a too so that

$$\{a\} \subset U \cap V, \text{ i.e. } U \cap V \neq \varnothing.$$

Consequently, X cannot be Hausdorff.

(3) Remember that

$$x \in \{a\}' \iff \forall U \in \mathcal{V}(x): \ U \cap \{a\} - \{x\} \neq \varnothing.$$

Since U contains a, we see immediately that only a does not belong to $\{a\}'$ and hence $\{a\}' = X - \{a\}$.

(4) Let $U \neq \varnothing$ be an open set in T. Then $a \in U$ and hence

$$\{a\} \subset U \Rightarrow \{a\}' \subset U' \Leftrightarrow X - \{a\} \subset U' \subset X.$$

Two cases are to be discussed:

(a) If $U' = X$, then $\overline{U} = U' \cup U = X \cup U = X$.

(b) If $U' = X - \{a\}$, then $\overline{U} = U' \cup U = X - \{a\} \cup U = X$ as $a \in U$.

Thus in either case open sets are dense in X.

(5) As for $\{a\}'$, we find that $A' = X - \{a\}$ for any set A that contains a.

(6) Yes X is separable for $\{a\}$ is a countable (finite!) subset of X that is dense by the last but one question because $\{a\}$ is open in T.

Now, $X - \{a\}$ is not separable. Assume it were and let $A \subset X - \{a\}$. Then A would be dense. But $a \notin A$ and $a \notin A' = X - \{a\}$ and hence $a \notin \overline{A}$. Thus $\overline{A} = A \cup A'$ would have to be equal to $X - \{a\}$, a clear contradiction. Therefore, $X - \{a\}$ is not separable.

(7) Let A be a proper subset of X. We have to show that $\overset{\circ}{\overline{A}} = \varnothing$. Since $\overline{A}$ is closed, $a \notin \overline{A}$. The biggest *open* set contained in $\overline{A}$ must contain $\{a\}$. Thus $\overset{\circ}{\overline{A}} = \varnothing$.

(8) The induced topology, as the interested reader may verify, is the discrete one and it is of course Hausdorff.

SOLUTION 3.3.26.

(1) Let us show that T is a topology on $[-a, a]$. First $\varnothing \in T$ since $\{0\} \not\subset \varnothing$ and $X = [-a, a] \in T$ since $(-a, a) \subset X = [-a, a]$.

Now let U and V be both in T and hence $(\{0\} \not\subset U$ or $(-a, a) \subset U)$ and $(\{0\} \not\subset V$ or $(-a, a) \subset V)$. In all possible cases we will have $U \cap V \in T$.

Finally, let $\{U_i\}_{i \in I}$ be an arbitrary collection of elements of T. If $\{0\} \not\subset U_i$ for all $i \in I$, then $\{0\} \not\subset \bigcup_{i \in I} U_i$ and hence $\bigcup_{i \in I} U_i \in T$. If at least one U_j does not contain $\{0\}$, then it will contain $(-a, a)$ and hence $(-a, a) \subset U_j \subset \bigcup_{i \in I} U_i$ and this also means that $\bigcup_{i \in I} U_i \in T$.

(2) The closed sets in this topology are $\{a\}$, $\{-a\}$, $\{-a, a\}$, $\varnothing$, $[-a, a]$ and any subset of $[-a, a]$ containing 0.

(3) The set $A = \{\frac{a}{3}\}$ is not closed, but from the previous question, $\{0, \frac{a}{3}\}$ is a closed set and it is clearly the smallest set which contains A. Thus

$$\overline{A} = \left\{0, \frac{a}{3}\right\}.$$

(4) Let B be any subset of $(-a, a) \subset X$. Let us show that 0 is a limit point for B, i.e. $0 \in B'$. We recall that

$$0 \in B' \iff \forall U \in \mathcal{V}(0) : U \cap B - \{0\} \neq \varnothing.$$

Since U is an open neighborhood of 0, we must have $(-a, a) \subset U$. Whence

$$U \cap B - \{0\} \neq \varnothing,$$

as required.

SOLUTION 3.3.27.

(1) First, $\varnothing \in T$ since $\exists a = 0 \in [0, 2]$ such that $\varnothing = [0, 0)$ and $X \in T$ since $\exists a = 2 \in [0, 2]$ such that $X = [0, 2)$. Now let $[0, a_i)_{i \in I}$ be an arbitrary collection of elements in T. Then

$$\bigcup_{i \in I} [0, a_i) = [0, a) \text{ where } a = \sup_{i \in I} a_i \in [0, 2].$$

This means that $\bigcup_{i \in I} [0, a_i) \in T$. Lastly, let $[0, a)$ and $[0, b)$ be two elements of T where $0 \le a \le 2$ and $0 \le b \le 2$. Then one has

$$[0, a) \cap [0, b) = [0, c) \text{ where } c = \min(a, b) \in [0, 2]$$

and so $[0, a) \cap [0, b) \in T$.

(2) One example among many is the following: Take $U_n = \left[0, \frac{2}{n}\right)$ for $n \ge 1$. It belongs to T for all $n \ge 1$. However, $\bigcap_{n=1}^{\infty} \left[0, \frac{2}{n}\right) = \{0\}$ which cannot be an element of T as it cannot be written in the form $[0, a)$ for any $a \in [0, 2]$.

(3) This topology cannot be separated as any two open sets will both contain zero and hence their intersection will never be empty.

(4) The closed sets are of the form $[b, 2)$ where $0 \le b \le 2$ (why?).

(5) There is somehow a technical way of answering this question but there is a more direct way of answering it. The closure of $A = \left[1, \frac{3}{2}\right]$ is by definition the smallest closed set containing A. But we have just seen that closed sets in X are of the form $[b, 2)$ where $0 \le b \le 2$. Therefore the closure of A is $[1, 2)$.

The interior of A, i.e. the largest open set contained in A, is empty as open sets are of the form $[0, a)$ where $0 \le a \le 2$ and none of them can be contained in A.

(6) Let $B = \mathbb{Q} \cap [0, 2)$. It is obviously countable. Besides, its closure, the smallest set containing it, is of the form $[b, 2)$ where $0 \le b \le 2$. Hence

$$\overline{\mathbb{Q} \cap [0, 2)} = [0, 2),$$

proving the density of B in (X, T). Thus X is separable.

SOLUTION 3.3.28.

(1) Let $x \in X$. We only show that $\overline{\{x\}} \subset \{x\}$ or equivalently $\{x\}^c \subset \overline{\{x\}}^c$. Let $y \in \{x\}^c$, i.e. $y \notin \{x\}$ or $y \neq x$. Since T is Hausdorff,

$$\exists (U, V) \in \mathcal{V}(x) \times \mathcal{V}(y) : \ U \cap V = \varnothing.$$

This gives us $U \cap \{x\} = \varnothing$. Thus $y \notin \overline{\{x\}}$, i.e. $y \in \overline{\{x\}}^c$.

(2) The result is no longer true if T is not assumed to be Hausdorff. Consider the topology T, on the set $X = \{1, 2, 3\}$, defined by

$$T = \{\varnothing, \{2\}, \{1, 2\}, \{2, 3\}, X\}.$$

It is plain that $\{2\}$ is not closed as $\{2\}^c = \{1, 3\}$ is not open. On can also check that T is not Hausdorff.

(3) Well, an example of that is the co-finite topology (Exercise 3.3.22) in which all finite sets, and in particular singletons, are closed.

(4) Let U_x be an open set containing x. We need to establish that $\bigcap_{x \in X} U_x = \{x\}$. Since for all $x \in X$, $x \in U_x$, we immediately see that $\{x\} \subset \bigcap_{x \in X} U_x$.

Conversely, let $y \in \bigcap_{x \in X} U_x$ with $y \neq x$. Since X is Hausdorff, for some open set U containing x and for some open set V containing y we have $U \cap V = \varnothing$. But y is in all open sets which contain x and hence $U \cap V \neq \varnothing$ which is a contradiction and so $y = x$. The proof is complete.

(5) Let X be $\mathbb{R}$ equipped with the co-finite topology. That X is *not* Hausdorff was already established in Exercise 3.3.22. We show that the intersection of all open sets containing x is $\{x\}$ itself. We only show $\bigcap_{x \in U} U \subset \{x\}$ where U is open in X and contains x. Let $y \in \bigcap_{x \in U} U$ and assume $y \neq x$. Then for all $U \in X$, we have $y \in U$. Hence $\mathbb{R} - \{y\}$, being an open set (why?) that contains x, would have to contain y too which is absurd! Thus $y = x$ and this completes the proof.

SOLUTION 3.3.29.

(1) $\mathbb{R}$ is obviously a union of elements of $\mathcal{B}$. An intersection of two elements of $\mathcal{B}$ is again an element of $\mathcal{B}$ for it is either the

empty set or an interval of the same type as the elements of $\mathcal{B}$. Therefore, $\mathcal{B}$ is a base for $\mathbb{R}$.

(2) $\mathbb{R}_\ell$ is strictly finer than $\mathbb{R}$. We have to show that any open set in $\mathbb{R}$ is an open set in $\mathbb{R}_\ell$ or in terms of bases, any basis element in $\mathbb{R}$, i.e. an open interval, can be written as a union of members of the basis of $\mathbb{R}_\ell$. This is illustrated in

$$\forall a, b \in \mathbb{R}, \ (a, b) = \bigcup_{n \in \mathbb{N}} \left[a + \frac{1}{n}, b \right).$$

To finish the proof, we need to exhibit an open set in $\mathbb{R}_\ell$ which is not one in $\mathbb{R}$. One choice is $[0, 1)$ and the proof is complete.

(3) First, from the previous question all open sets in $\mathbb{R}$ are open in $\mathbb{R}_\ell$. The same thing for closed sets. There are many other sets that are open and/or closed in $\mathbb{R}_\ell$. For example,

$$(-\infty, b) = \bigcup_{n \in \mathbb{N}} [-n, b), \ [a, \infty) = \bigcup_{n \in \mathbb{N}} [a, n)$$

are open. They are also closed as

$$(-\infty, b)^c = [b, \infty) \text{ and } [a, \infty)^c = (-\infty, a)$$

are open. Also $[a, b)$ is open and it is also closed as

$$[a, b)^c = (-\infty, a) \cup [b, \infty)$$

is open.

There are of course non-clopen sets. For instance, $[a, b]$ is closed but not open (why?), and (a, ∞) is open and not closed. Finally, there are sets which are neither open nor closed like $(a, b]$ (prove it!).

(4) Yes, $\mathbb{R}_\ell$ is separated since $\mathbb{R}_\ell$ is finer than $\mathbb{R}$. This was discussed and proved in the section "True or False" of this chapter.

We propose another method to show that $\mathbb{R}_\ell$ is not Hausdorff. Let $x \neq y$ be two distinct reals. Take $x < y$ for example. Then it is clear that $[x, y)$ is an open set in $\mathbb{R}_\ell$ containing x and $[y, y + 1)$ is an open set in $\mathbb{R}_\ell$ containing y. Besides, these two sets obey

$$[x, y) \cap [y, y + 1) = \varnothing,$$

completing the proof.

(5) Yes, $\mathbb{R}_\ell$ is separable. To see that, first note that $\mathbb{Q}$ is dense in $\mathbb{R}$ (this is independent of topology!). To prove the density of $\mathbb{Q}$ in $\mathbb{R}_\ell$, note that unions of intervals of the form $[a, b)$ always intersect $\mathbb{Q}$.

SOLUTION 3.3.30.

(1) $\mathbb{R}_K$ is finer than $\mathbb{R}$ since its basis contains the basis of $\mathbb{R}$. It is strictly finer since $\mathbb{R} - K$ is open in $\mathbb{R}_K$ (it is a union of sets of the form $(a, b) - K$) but it is not open in $\mathbb{R}$ for K is not closed.

(2) As in the previous exercise, $\mathbb{R}_K$ is Hausdorff as it is finer than $\mathbb{R}$.

(3) No. The two topologies are not comparable and it is better to use bases elements.
   (a) $\mathbb{R}_K \not\subset \mathbb{R}_\ell$: For $(-1, 1) - K$ is open in $\mathbb{R}_K$ (clear!) but not in $\mathbb{R}_\ell$.
   (b) $\mathbb{R}_\ell \not\subset \mathbb{R}_K$: For $[-1, 0)$ is in $\mathbb{R}_\ell$ but not in $\mathbb{R}_K$.

(4) Yes, K is closed in $\mathbb{R}_K$ for its complement is $\mathbb{R} - K$, hence it can be written as $\bigcup_{a,b\in\mathbb{R}} ([a, b) - K)$ which is open.

(5) We claim that $K' = \varnothing$. Any point in K is not a limit point for K (why?).

   If $x \notin K$, then $U = \mathbb{R} - K$ is an open set containing x and verifying $U \cap K - \{x\} = \varnothing$. Thus $K' = \varnothing$.

SOLUTION 3.3.31.

(1) Let $x \in \mathbb{R}$. Then there are always $a, b \in \mathbb{Q}$ such that $x \in (a, b)$ (why?). Now, let (a, b) and (c, d) be in $\mathcal{B}$ (a, b, c and d are tacitely assumed to be rationals!). If x is in the intersection of these sets, then $(a, b) \cap (c, d)$ is an interval of the same type. This proves that $\mathcal{B}$ is a basis.

   Now, we prove $\mathcal{B}$ actually generates the usual topology of $\mathbb{R}$. Let U be an open set in $\mathbb{R}$. Let $x \in U$. Then

$$\exists y, z \in \mathbb{R} : x \in (y, z) \in U.$$

By the density of $\mathbb{Q}$ in $\mathbb{R}$,

$$\exists a, b \in \mathbb{Q}, \ a \in (y, x) \text{ and } b \in (x, z)$$

leading to

$$x \in (a, b) \subset U \ (a, b \in \mathbb{Q}).$$

The proof is over.

(2) For any x real, there are rationals a and b verifying $a \leq a < x < b$. Now, if the intersection of two basis elements of $\mathcal{B}'$ is not empty, then it is necessarily of their form. Thus $\mathcal{B}'$ is a basis. Now, we have to show that $\mathbb{R}_{\mathcal{B}'} \neq \mathbb{R}_\ell$. It is clear that

$[\pi, 4) \in \mathbb{R}_\ell$. If $[\pi, 4)$ were in $\mathbb{R}_{\mathcal{B}'}$, then there would be some rationals a and b such that

$$\pi \in [a, b) \subset [\pi, 4).$$

It then becomes clear that no rational a would satisfy that condition. Therefore, $\mathcal{B}'$ does not generate the lower limit topology on $\mathbb{R}$.

SOLUTION 3.3.32. Remember that in usual $\mathbb{R}$,

$$d(A) = \sup_{x,y \in A} |x - y|$$

where $A \subset \mathbb{R}$. We then easily find that

$$d((0,1) \cap \mathbb{Q}) = d((0,1) \cap \mathbb{R} \setminus \mathbb{Q}) = 1.$$

SOLUTION 3.3.33.

(1) The set $\{d(x, a) : a \in A\}$ is obviously non-empty. It is also bounded from below as

$$d(x, a) \geq 0, \ \forall a \in A.$$

Thus $d(x, A)$ exists.

(2) Let $\varepsilon > 0$ and let $B(x, \varepsilon)$ be the open ball of center x and radius ε. We then have

$$\begin{aligned}
x \in \overline{A} &\Longleftrightarrow \forall \varepsilon > 0, B(x, \varepsilon) \cap A \neq \varnothing \\
&\Longleftrightarrow \forall \varepsilon > 0, \exists a_\varepsilon \in A : \ d(x, a) < \varepsilon \\
&\Longleftrightarrow \forall \varepsilon > 0, \exists a_\varepsilon \in A : \ 0 \leq d(x, a) < \varepsilon + 0 \\
&\Longleftrightarrow d(x, A) = 0
\end{aligned}$$

by the greatest lower bound property.

Let $a \in X$. We have to show that $\{a\}$ is closed. We have

$$\overline{\{a\}} = \{x \in X : \ d(x, a) = 0\} = \{a\},$$

i.e. $\{a\}$ is closed.

REMARK. This is a particular case and another proof of the result of Exercise 3.3.28 since every metric space is Hausdorff.

(3) Since $A \subset \overline{A}$, we have

$$d(x, A) \geq d(x, \overline{A}).$$

Now for all $a \in A$ and for all $b \in \overline{A}$,

$$d(x, A) \leq d(x, a) \leq d(x, b) + d(b, a).$$

Hence
$$d(x, A) \leq d(x, b) + d(b, A).$$
But, $d(b, A) = 0$ as $b \in \overline{A}$ and so
$$d(x, A) \leq d(x, b), \ \forall b \in \overline{A}.$$
Taking the inf again over $b \in \overline{A}$ gives
$$d(x, A) \leq d(x, \overline{A})$$
leading to the wanted equality.

REMARK. The result of Question 3 can be interpreted as follows: A point in the closure of A is not very far from A.

SOLUTION 3.3.34.

(1) First, note that $A \times B \subset \overline{A} \times \overline{B}$. But $\overline{A} \times \overline{B}$ is closed (why?) and hence $\overline{A \times B} \subset \overline{A} \times \overline{B}$.

Now we prove the other inclusion. Let $(x, y) \in \overline{A} \times \overline{B}$. Then $x \in \overline{A}$ and $y \in \overline{B}$. Hence any neighborhood of x intersect A and so does any neighborhood of y with B. Let Ω be a neighborhood of (x, y). Then Ω is a union of elements of the form $U \times V$ where U is open in X and contains x, V is open in Y and contains y. We have
$$(U \times V) \cap (A \times B) = (U \cap A) \times (V \cap B) \neq \varnothing$$
as $x \in \overline{A}$ and $y \in \overline{B}$. Thus Ω too intersects $A \times B$ which leads to $(x, y) \in \overline{A \times B}$.

(2) There different methods to prove this property. The one we use here is based on the known property $\overline{A^c} = (\overset{\circ}{A})^c$ and on the previous question. We have
$$\begin{aligned}
(\overset{\circ}{\overbrace{A \times B}})^c = \overline{(A \times B)^c} &= \overline{(A^c \times Y) \cup (X \times B^c)} \\
&= \overline{A^c \times Y} \cup \overline{X \times B^c} \\
&= (\overline{A^c} \times \overline{Y}) \cup (\overline{X} \times \overline{B^c}) \\
&= (\overline{A^c} \times Y) \cup (X \times \overline{B^c}) \\
&= ((\overset{\circ}{A})^c \times Y) \cup (X \times (\overset{\circ}{B})^c) \\
&= (\overset{\circ}{A} \times \overset{\circ}{B})^c.
\end{aligned}$$

Thus
$$\overset{\circ}{\overbrace{A \times B}} = \overset{\circ}{A} \times \overset{\circ}{B}.$$

SOLUTION 3.3.35. Using a proof by induction, it suffices to prove this result for two spaces. Let X and Y be two separable spaces, i.e. there are two *countable* subsets A and B of X and Y respectively such that

$$\overline{A} = X \text{ and } \overline{B} = Y.$$

Now, obviously $A \times B$ is countable. It is also dense in $X \times Y$ since

$$\overline{A \times B} = \overline{A} \times \overline{B} = X \times Y.$$

The solution is complete.

SOLUTION 3.3.36.

(1) A has an empty interior since no open ball can be contained in A (why?). Its closure is given by

$$\overline{A} = A \cup \{(0, y) : \ -1 \le y \le 1\}.$$

This comes from the fact any open ball centered at $(0, y)$ with $-1 \le y \le 1$ intersects A.

(2) The interior of B is void. For no open ball can be contained in A regardless of its radius. Also B is closed. There are different ways of seeing this:

    (a) Let $(x, y) \notin A$. Then there is always some $r > 0$ such that $B((x, y), r)$ does not intersect B and hence $(x, y) \notin \overline{B}$. The proof is over.

    (b) Alternatively and anticipating a result on continuity (to be seen in the next chapter) the given set B is the graph of the function $x \mapsto x$ which is continuous on usual $\mathbb{R}$ and the graph of a continuous function is closed.

    (c) Also, B is the diagonal of $\mathbb{R}$ and it is closed since usual $\mathbb{R}$ is Hausdorff (cf Exercise 4.3.31).

(3) The answer becomes obvious once we write $C = C_1 \times C_2$ where

$$C_1 = (-2, 2) \text{ and } C_2 = (-3, 3).$$

Hence

$$\overset{\circ}{C} = \overset{\circ}{\overbrace{C_1 \times C_2}} = \overset{\circ}{C_1} \times \overset{\circ}{C_2} = (-2, 2) \times (-3, 3) = C.$$

Similarly

$$\overline{C} = \overline{C_1} \times \overline{C_2} = [-2, 2] \times [-3, 3].$$

(4) Since $\{(1, 1)\}$ is a singleton in usual $\mathbb{R}^2$, we immediately get

$$\overset{\circ}{D} = \overset{\circ}{\overbrace{\{(1,1)\} \times C}} = \{(1,1)\} \times \overset{\circ}{\overbrace{C}} = \varnothing \times \overset{\circ}{C} = \varnothing.$$

Finally,

$$\overline{D} = \overline{\{(1,1)\}} \times \overline{C} = \{(1,1)\} \times [-2,2] \times [-3,3].$$

SOLUTION 3.3.37. Let $\varphi : X \to X/R$ be the quotient map. Let

$$T = \{A \in X/R : \ \varphi^{-1}(A) \text{ is open in } X\}.$$

Let us show that T is a topology in X/R.

(1) $\varnothing, X/R \in T$ as: $\varphi^{-1}(\varnothing) = \varnothing$ and $\varphi^{-1}(X/R) = X$.

(2) Let $A, B \in T$. Then

$$\varphi^{-1}(A \cap B) = \varphi^{-1}(A) \cap \varphi^{-1}(B)$$

is open in X so that $A \cap B \in T$.

(3) Let $(A_i)_{i \in I}$ be a collection of elements in T. Then

$$\varphi^{-1}(\bigcup_{i \in I} A_i) = \bigcup_{i \in I} \varphi^{-1}(A_i)$$

is open in X and so $\cup_{i \in I} A_i \in T$.

SOLUTION 3.3.38.

(1) Left to the reader!

(2) Denote the quotient map by p, i.e. the map $p : \mathbb{R} \to \mathbb{R}/\mathbb{Q}$. Let $[s]$ and $[r]$ be two elements of $\mathbb{R}/\mathbb{Q}$. Let $U \in \mathcal{V}([s])$ and $V \in \mathcal{V}([r])$. By definition, $p^{-1}(U)$ and $p^{-1}(V)$ are two open sets in $\mathbb{R}$. Hence

$$\exists q, q' \in \mathbb{Q} : \ q \in p^{-1}(U), q' \in p^{-1}(U) \ \text{(why?)}$$

Thus $[q] \in U$ and $[q'] \in V$. But $[q] = [q']$ since $q - q' \in \mathbb{Q}$. Therefore, $U \cap V$ is never empty and consequently $\mathbb{R}/\mathbb{Q}$ is not Hausdorff.

(3) Let U be any non-empty set in $\mathbb{R}/\mathbb{Q}$. We must show that $U = \mathbb{R}/\mathbb{Q}$. Denote the quotient map by p. Then $p^{-1}(U)$ is open in $\mathbb{R}$. Now, the map $t \mapsto t + \alpha$ is continuous for each real α. Whence the set $\{t \in \mathbb{R} : \ t + \alpha \in p^{-1}(U)\}$ is open in $\mathbb{R}$. Thus it must necessarily intersect $\mathbb{Q}$ and so

$$\exists q \in \mathbb{Q} : \ q + \alpha \in p^{-1}(U).$$

But $p(\alpha) = p(q + \alpha)$ (why?). This implies that $p(\alpha) \in U$, i.e. $\alpha \in p^{-1}(U)$ and hence $p^{-1}(U) = \mathbb{R}$. Therefore, $U = \mathbb{R}/\mathbb{Q}$. The proof is complete.

3.4. Hints/Answers to Tests

SOLUTION 7.

(1) We can write (can't we?) $A = \bigcup_{x \in A} U_x...$

(2) It reminds us of the definition of an open set in a metric space with U_x playing the role of an open ball...

SOLUTION 8. Yes! why?...

SOLUTION 9. No! Consider e.g. $\{\pi\}$...

SOLUTION 10. No! (why?)...

SOLUTION 11. No T is not a topology on X. While $\varnothing$ and X both belongs to T (why?), the union of two elements in T need not remain in T (consider A_4 and A_3 for example)...

SOLUTION 12. There are nine topologies having four open sets. Find them directly or just apply Exercise 3.5.1.

SOLUTION 13. Construct such a set using sets similar to $\{\frac{1}{n} : n \in \mathbb{N}\}$ which we know it has 0 as its *unique* limit point. What is left to do should be clear to the reader by now...

SOLUTION 14.

(1) In the discrete topology, $A' = \varnothing$. $\{x\}$ is an open set containing x...

(2) In the co-finite topology, $A' = X$ if A is infinite and $A' = \varnothing$ if A is finite. The reason is that if A is infinite, then $X - \{a\}$ is open and contains x where $a \in X$...and if A is a singleton, consisted of the element a say, then $X - \{a\}$ is always open and does not intersect A and similar arguments work for A consisted of a finite number of elements...

SOLUTION 15. No. Why?...

SOLUTION 16. Let T be the usual topology on $\mathbb{R}$ and let T' be the co-countable topology on $\mathbb{R}$. Then T is stronger than T' for it has more closed sets as there are closed sets which are uncountable...

SOLUTION 17. The answer is no as every proper subset in this topology is closed...

SOLUTION 18.

(1) The empty set corresponds to the case $a = 0$. The rest is obvious too...

(2) Closed sets are of the form $(-\infty, -a] \cup [a, +\infty)$. For $[-1, 2]$, the smallest closed superset is $\mathbb{R}$ and the largest open subset is $(-1, 1)$.

(3) The closure and the interior of $\{0\}$ are given by $\mathbb{R}$ and $\varnothing$ respectively.

As for $\{1\}$ they are given by $\mathbb{R} \setminus (-1, 1)$ and $\varnothing$ respectively.

SOLUTION 19. It inherits the discrete topology. Every singleton $\{x\}$ $(x \in A)$ is open in A as it can be written as $A \cap \{x, a\}$...

SOLUTION 20. Yes, if this set is clopen. Otherwise, this cannot occur as the frontier of a set is always *closed* (is it not?).

SOLUTION 21. Yes (why?)...

SOLUTION 22. Well, it is a routine so do it!...

CHAPTER 4

Continuity and Convergence

4.2. True or False: Answers

ANSWERS.

(1) No, this is not always the case if the topologies endowing the domain and the "arrival" sets are different. See Exercise 4.3.1. If, however, the identity mapping is between two identical spaces endowed with the same topologies, then it is continuous.

(2) It is asked whether each continuous function is open? Such is not the case. As a counterexample, let $f : \mathbb{R} \to \mathbb{R}$ defined by $f(x) = 0$ ($\mathbb{R}$ endowed with its standard topology). Then f is continuous but for some (and here any!) open U in $\mathbb{R}$, $f(U) = \{0\}$ is not open in $\mathbb{R}$.

(3) The left-to-right implication is correct and for a proof see Exercise 4.3.19. As for the backward implication, it is not true. For a counterexample, take the function

$$f(x) = \begin{cases} 1, & x \in \mathbb{Q}, \\ 0, & x \notin \mathbb{Q}. \end{cases}$$

Then f is not continuous whilst $f_{\mathbb{Q}}$ is continuous (both in the usual topology).

(4) In general, only the right-to-left implication is verified. To see this, let U be an open set containing x. Since (x_n) converges to x, for any open set containing x, and in particular for U,

$$\exists N \in \mathbb{N}, \forall n \in \mathbb{N} \ (n \geq N \implies x_n \in U)$$

and hence $A \cap U \neq \varnothing$ or $x \in \overline{A}$.

The other implication may fail to hold. In $\mathbb{R}$ equipped with the co-countable topology, let $A = [0, 2]$. Then $\overline{A} = \mathbb{R}$ (cf. Exercise 3.3.24). Now, the only convergent sequences in this space are the eventually constant ones (see Exercise 4.3.14). Hence

$$\forall x_n \in [0, 2], \ x_n \nrightarrow -1 \text{ and yet } -1 \in \overline{A}.$$

If, however, we are dealing with metric spaces only, then the equivalence always always holds and it is a very useful result to use.

Let us then show the left-to-right implication in metric spaces. Let X be endowed with a metric d and let $x \in \overline{A}$. Then

$$\forall \varepsilon > 0 : \ B(x, \varepsilon) \cap A \neq \varnothing$$

and hence

$$\forall n \in \mathbb{N} : \ B\left(x, \frac{1}{n}\right) \cap A \neq \varnothing.$$

Choosing an x_n in this intersection gives us for all n, $d(x_n, x) < \frac{1}{n}$ (and $x_n \in A$). Therefore, (x_n) converges to x in (X, d).

(5) This statement as it stands is something to avoid imperatively in topology. One has to be more precise about the space in which the convergence is to be established. For instance, in the usual topology of $\mathbb{R}$, this sequence converges to zero while in other spaces (or/and topologies) it can have different limits (see Exercise 4.3.12).

(6) The reasoning has a problem with the passage $\frac{1}{n} \not\subset U$ implying $\frac{1}{n} \in U^c$. This is wrong since $\frac{1}{n}$ not being in U means that $(\frac{1}{n}) \not\subset U$ and this does not imply necessarily that $(\frac{1}{n}) \subset U^c$. For one correct proof see Exercise 4.3.12.

(7) First, A is closed as its complement, being an arbitrary union of open sets, is open!

Second, the known result says that a set is closed in a metric space, if whenever a sequence in this set converges, then it must have a limit inside that set. In our case, that result cannot be applied as (x_n) does not even converge!

(8) A priori, the reader might think something is wrong but in fact everything is fine and nothing contradicts the continuity of f. First, f is in effect continuous and this only requires basic one variable analysis. Second, $[0, 1]$ is well closed in $\mathbb{R}$ and its preimage, given by

$$f^{-1}([0, 1]) = \left\{x \in \mathbb{R}_+^* : \ \ln x \in (0, 1)\right\} = (1, e],$$

is not closed in $\mathbb{R}$ but it is *closed* in the subspace topology of $\mathbb{R}^*$! (the topology which should be used here).

(9) The answer is no! For counterexamples, see Exercise 4.3.13.

(10) Only the left-to-right implication is true. In other words, the uniqueness of the limit of a sequence does not characterize the Hausdorffness property. We give a proof. Assume a given

sequence (x_n) in a separated space X has two different limits, x and y say. Since X is Hausdorff,

$$\exists U \in \mathcal{V}(x), \exists V \in \mathcal{V}(y) : \ U \cap V = \varnothing.$$

Since (x_n) converges to a, for all open sets containing x and in particular for U

$$\exists N_1 \in \mathbb{N}, \ \forall n \in \mathbb{N} : \ (n \geq N_1 \Rightarrow x_n \in U).$$

Similarly,

$$\exists N_2 \in \mathbb{N}, \ \forall n \in \mathbb{N} : \ (n \geq N_2 \Rightarrow x_n \in V).$$

So, for $n \geq \max(N_1, N_2)$, $x_n \in U \cap V$ which contradicts the fact that U and V are disjoint! Thus the limit is unique.

As for the other implication we present a counterexample. Consider $X = \mathbb{R}$ equipped with the co-countable topology. Then the only convergent sequences are the eventually constant ones (see Exercise 4.3.14). Let (x_n) be a sequence in X which converges to two different limits, x and y $(x \neq y)$, say. Then

$$\exists N_1 \in \mathbb{N}, \ \forall n \geq N_1 : \ x_n = a \text{ and } \exists N_2 \in \mathbb{N}, \ \forall n \geq N_2 : \ x_n = b.$$

Hence for $n \geq \max(N_1, N_2)$ we would have $x_n = a = b$ which contradicts the hypothesis $a \neq b$. Thus the limit is unique. However, we already know from Exercise 3.3.24 that X is not Hausdorff.

(11) The answer is yes but some comments have to be given. First, an open set U in $\mathbb{R}$ is written as

$$U = \bigcup_{i \in I} (a_i, b_i).$$

Then

$$f^{-1}(U) = f^{-1}\left(\bigcup_{i \in I} (a_i, b_i) \right) = \bigcup_{i \in I} f^{-1}(a_i, b_i).$$

Since the arbitrary union of open sets is open, it suffices to have $f^{-1}(a_i, b_i)$ open which is the hypothesis.

We also observe that this is true since $\{(a_i, b_i)\}_{i \in I}$ is a basis in $\mathbb{R}$. Hence if B_i is some basis in some topological space X and $f : Y \to X$ is a function (Y is a topological space), then it is sufficient to check the openness of $f^{-1}(B_i)$ in Y to establish the continuity of f. This will occur from time to time in the sequel.

(12) Two topological spaces (X and Y say) are said to be homeomorphic if there is a homeomorphism $f : X \to Y$ (or obviously
$f : Y \to X$). Thus it becomes apparent that it is quite easy
to show that two spaces are homeomorphic since it suffices for
that purpose to exhibit *one* homeomorphism between the two
spaces.

However, if one wants to show that two spaces X and Y
are not homeomorphic, then one sees immediately that it is
not an easy matter as one will have to show that there is
no homoeomorphism between the two spaces. An advanced
topology course, namely algebraic topology, is a powerful tool
for proving that two spaces are not homeomorphic. This is
not discussed in this book. However, in Chapters 5 and 6,
some criteria will be used to prove that some spaces are not
homeomorphic.

(13) Since f is continuous, we have $f(\overline{A}) \subset \overline{f(A)}$. Since A is dense
in X, we have $\overline{A} = X$. Then

$$f(X) = f(\overline{A}) \subset \overline{f(A)}.$$

Hence, the closure of $f(A)$ in $f(X)$, given by $f(X) \cap \overline{f(A)}$, is
equal to $f(X)$. Thus $f(A)$ is dense in $f(X)$.

(14) True! To see this, let $f : X \to Y$ be a homeomorphism between two topological spaces where X is separable. Let us
show that Y is separable. Since X is separable, there exists a
countable subset A such that $\overline{A} = X$. By the previous answer
$f(A)$ is dense in $f(X) = Y$. But

$$f(A) = \{f(x) : \ x \in A\}$$

is obviously countable. The proof is over.

(15) The answer is again no. One has to distinguish between an algebraic notion and a topological one. The bijectivity is purely
algebraic whilst the continuity is topological. There are many
counterexamples which the reader will see below.

(16) No! Consider $f(x) = x$ from $X = \mathbb{R}$ (endowed with the usual
topology) onto $Y = \mathbb{R}$ (endowed with the discrete topology).
Then f is bijective. It is also open and closed because every
subset of Y is open and closed. It is, however, not continuous
since $\{2\}$ is open in Y but its preimage (itself in this case) is
not open in X.

(17) The answer is yes. First, the known result states that f is
continuous iff $f(\overline{A}) \subset \overline{f(A)}$. Since f is already bijective, it
only remains to check that f^{-1} is continuous iff $f(\overline{A}) \supset \overline{f(A)}$.

Let $B \subset Y = f(X)$, then for some $A \subset X$: $f(A) = B$ and hence $A = f^{-1}(B)$. Then

$$f(\overline{A}) \supset \overline{f(A)} \Leftrightarrow f\left(\overline{f^{-1}(B)}\right) \supset \overline{B} \Leftrightarrow \overline{f^{-1}(B)} \supset f^{-1}(\overline{B}),$$

that is, if and only if f^{-1} is continuous.

(18) The answer is no! We give a counterexample. Let $Y = \mathbb{R}$ endowed with the usual topology and let $X = \mathbb{R}$ be endowed with the discrete topology. Let $f : X \to Y$ defined for all $x \in \mathbb{R}$ by $f(x) = x$. Then f is a bijection. Then f is continuous as for every open U set in Y, $f^{-1}(U)$ is open in X. However, its inverse, i.e. $f^{-1} : Y \to X$ is not continuous (why?).

The reader *must not* think this is solely true with different topologies or that this cannot occur in the usual topology. For instance, let $f : X = [0,1) \cup \{3\} \to Y = [0,1]$ be defined for all $x \in X$ by

$$f(x) = \begin{cases} x, & 0 \le x < 1, \\ 1, & x = 3 \end{cases}$$

and both X and Y are endowed with the induced usual topology of $\mathbb{R}$. Details are left to the reader.

(19) The answer is yes. To see this let $f : X \to Y$ be a function with the listed properties. We have to show that $f^{-1} : Y \to X$ is continuous, i.e. for every open set U in X, $(f^{-1})^{-1}(U)$ is open in Y. But since f is bijective, we have

$$(f^{-1})^{-1}(U) = f(U)$$

which is open by the openness of f. Thus f^{-1} is continuous and hence f is a homeomorphism.

(20) The answer is yes. For a proof see Exercise 4.3.6.

(21) False! Let $f : (0,1) \to \{1\}$ be the constant function (both sets with respect to the usual topology). Then from classical analysis, f is continuous. Then 0 is a limit point for $(0,1)$ while $f(0) = 1$ is not a limit point for $\{1\}$.

(22) False! Only the left-to-right implication holds. For a proof and for a counterexample to the other implication, see Exercise 4.3.33.

The right-to-left implication holds if X and Y are *Banach* spaces and if f is *linear*. This implication with the quoted hypotheses (and a fortiori, the whole equivalence) is a very important result in functional analysis called the "**Closed Graph Theorem**".

See Exercise 5.5.9 for another result.

4.3. Solutions to Exercises

SOLUTION 4.3.1.

(1) (a) If card$X \geq 2$, then the given function is not continuous. For if U is an open (different from $\varnothing$ and X) set in Y, then $f^{-1}(U) = U$ is not open in X as the only open sets in X are $\varnothing$ and X.

  (b) Now if card$X = 1$, then f is obviously continuous as the discrete and indiscrete topologies coincide in this case.

(2) No f is not continuous. For example, $\{0\}$ is open in Y but its preimage $\{1\}$ is not open in X.

(3) Let $U = (0, 1)$ be an open in Y. Then

$$f^{-1}(U) = \{x \in \mathbb{R} : \; x^2 \in (0, 1)\} = (-1, 0) \cup (0, 1)$$

which is not open in X as it is not of the form $\varnothing$ or $\mathbb{R}$ or $(a, +\infty)$.

(4) The function f in this case is continuous since for any open set U in Y, $f^{-1}(U)$ is open since it is a subset of X. This means that if $X = [0, 3]$, say, is given the discrete topology, then a function like

$$x \mapsto f(x) = \begin{cases} -1, & 0 \leq x < 1, \\ 0, & 1 \leq x < 2, \\ 2, & 2 \leq x < 3, \end{cases}$$

will be continuous. This type of functions (i.e., those defined on a discrete topology) will have little interest in practise, but it is quite enlightening as a source of counterexamples.

(5) Let $f(x) = b$ for all $x \in X$. Let U be open in Y. We have

$$f^{-1}(U) = \{x \in X : \; f(x) \in U\} = \{x \in X : \; b \in U\} = \varnothing \text{ or } X$$

depending on whether $b \notin U$ or $b \in U$. Anyway, in either case $f^{-1}(U)$ is open in X and hence f is continuous.

(6) Take $f : A \to X$, where $A \subset X$, such that $f(x) = x$ for all $x \in A$. For any open U in X, $f^{-1}(U) = U \cap A$ which is open in A (in the subspace topology). Thus f is continuous.

SOLUTION 4.3.2. Remember that a function $f : X \to Y$ (X and Y being two topological spaces) is continuous at $x \in X$ if

$$\forall U \in \mathcal{V}(f(x)), \; f^{-1}(U) \in \mathcal{V}(x).$$

(1) f is not continuous at a for

$$\exists U = \{a, b\} \in \mathcal{V}(a) \text{ and } f^{-1}(U) = \{a, c\} \notin \mathcal{V}(a).$$

(2) f is continuous at b because for any open set containing $f(b) = c$ (in this case there is only one, namely X), $f^{-1}(X) = X$ is an open set that contains b!

(3) f is not continuous at c since

$$\exists U = \{b\} \in \mathcal{V}(f(c)) = \mathcal{V}(b) \text{ and } f^{-1}(U) = \{c\} \notin \mathcal{V}(b)$$

as $\{c\}$ is not open in X.

SOLUTION 4.3.3. Assume that $f^{-1}(\overset{\circ}{U}) \subset \overparen{f^{-1}(U)}$ holds for all U in Y. We must show that f is continuous. Let V be an open set in Y. Then $V = \overset{\circ}{V}$ and hence by hypothesis we obtain

$$\overparen{f^{-1}(V)} \subset f^{-1}(V) = f^{-1}(\overset{\circ}{V}) \subset \overparen{f^{-1}(V)}.$$

Therefore, $\overparen{f^{-1}(V)} = f^{-1}(V)$, proving that $f^{-1}(V)$ is open or that f is continuous.

Conversely, suppose f is continuous and let $U \subset Y$. Since $\overset{\circ}{U}$ is open, obviously so will be $f^{-1}(\overset{\circ}{U})$. Besides, $\overset{\circ}{U} \subset U$ and thus

$$f^{-1}(\overset{\circ}{U}) = \overparen{f^{-1}(\overset{\circ}{U})} \subset \overparen{f^{-1}(U)}.$$

The proof is complete.

SOLUTION 4.3.4. First, we note that f is obviously a bijection. The function $f : T' \to T$ is continuous. To see this, take any nonvoid (the case of an empty set trivially holds) open set U in T, then U^c is finite and hence it is countable, i.e. $U \in T'$. But $f^{-1}(U) = U$. Thus f is continuous. Since a countable set is not necessarily finite, we deduce that $f^{-1} : T \to T'$ is not continuous. Therefore, f is not a homeomorphism.

SOLUTION 4.3.5.

(1) The bijectivity of f is evident. Let us show that f is continuous. Since d and d' are topologically equivalent (see Exercise 2.3.27), f is a homeomorphism.

(2) If $X = \mathbb{R}$ and $d = |\cdot|$ the usual metric, then $(\mathbb{R}, d)$ is unbounded and $(\mathbb{R}, d')$ is bounded and yet these two spaces are homeomorphic.

SOLUTION 4.3.6. Let f be a homeomorphism between two topological spaces X and Y. Assume that X is Hausdorff, and let us show that Y is in its turn Hausdorff. Let $y, y' \in Y$ be such that $y \neq y'$. By the bijectivity of f, there exist *unique* $x, x' \in X$ such that $y = f(x)$ and $f(y') = x'$. By the bijectivity of f^{-1}, say, we see that x and x' must be *different*. By the Hausdorffness of X, we get

$$\exists (U, U') \in \mathcal{V}(x) \times \mathcal{V}(x') : U \cap U' = \varnothing.$$

Since f is open (and since U and U' are open), $f(U)$ and $f(U')$ are also open. They obviously contain y and y' respectively. But f is injective and hence

$$f(U) \cap f(U') = f(U \cap U') = \varnothing,$$

proving that $f(U)$ and $f(U')$ are disjoint. On that account, Y is Hausdorff.

SOLUTION 4.3.7.

(1) (a) Let (a, b) and (c, d) be any two intervals in $\mathbb{R}$. Define a function $f : (a, b) \to (c, d)$ defined by

$$f(x) = c + (d - c)\frac{x - a}{b - a}$$

for each $x \in (a, b)$. Then it is clear that f is continuous and bijective. Its inverse, $f^{-1} : (c, d) \to (a, b)$ given by

$$f^{-1}(x) = a + (b - a)\frac{x - c}{d - c}$$

for each $x \in (c, d)$, is obviously continuous too. Thus f is a homeomorphism and hence (a, b) and (c, d) are homeomorphic.

REMARK. Needless to recall that this is in the usual topology and that in other topologies these two intervals may not be homeomorphic.

(b) We leave it to the reader to check that $\mathbb{R}$ is homeomorphic to $(-1, 1)$ via the *homeomorphism*

$$f(x) = \frac{x}{1 + |x|}, \quad x \in \mathbb{R}.$$

Since the "homeomorphism relation" is transitive, then $\mathbb{R}$ is homeomorphic to any open interval by the previous question.

(2) The answer is yes! Remember that $\overline{\mathbb{R}} = \mathbb{R} \cup \{-\infty, +\infty\}$. For a possible homeomorphism consider

$$f(x) = \begin{cases} \frac{x}{1 + |x|}, & x \in \mathbb{R}, \\ -1, & x = -\infty, \\ 1, & x = \infty. \end{cases}$$

SOLUTION 4.3.8. This is easy. We have $A = f^{-1}(\{a\})$. Since $\{a\}$ is closed in $\mathbb{R}$, so is A since it is the preimage of a closed set under a continuous function.

SOLUTION 4.3.9.

(1) The function $(x, y) \mapsto f(x, y) = xy$ defined on $\mathbb{R}^2$ since it is a polynomial. Now A is closed as it is the inverse image of a closed set, that is $\{1\}$, by a continuous function.

(2) As before, the function $(x, y) \mapsto f(x, y) = x^2 + y^2$ defined on $\mathbb{R}^2$ is continuous since it is a polynomial. Hence

$$A = \{(x, y) \in \mathbb{R}^2 : x^2 + y^2 \leq 1\} = f^{-1}([0, 1])$$

is closed for $[0, 1]$ is closed in $\mathbb{R}$.

(3) First, this space has an algebraic dimension equal to n^2. The "function determinant" defined in $X = \mathcal{M}_n(\mathbb{R})$ and taking values in $\mathbb{R}$ is a polynomial of degree n^2 and hence it is continuous. The remaining part of the answer is a routine.

SOLUTION 4.3.10.

(1) The set A is closed since $A = f^{-1}((-\infty, a])$, f is continuous and $(-\infty, a]$ is closed in $\mathbb{R}$.

(2) The converse is not always true. Consider the *discontinuous* function (at $x = 0$)

$$f(x) = \begin{cases} 0, & x \leq 0 \\ 2, & x > 0 \end{cases}.$$

Now we show that for any a, the resulting set A will always be closed. We have
   (a) $a < 0 \Rightarrow A = \varnothing$, i.e. A is closed in $\mathbb{R}$.
   (b) $0 \leq a < 2 \Rightarrow A = (-\infty, 0]$, i.e. A is closed in $\mathbb{R}$.
   (c) $a \geq 2 \Rightarrow A = \mathbb{R}$, i.e. A is closed in $\mathbb{R}$.

SOLUTION 4.3.11. The left-to-right implication is evident. Let us prove the right-to-left implication. Since $\{(a, b) : a, b \in \mathbb{R}\}$ is a basis for usual $\mathbb{R}$, it suffices to prove that the inverse image of (a, b) via f is open. We have

$$(a, b) = (-\infty, b) \cap (a, \infty)$$

and hence

$$f^{-1}((a, b)) = f^{-1}((-\infty, b) \cap (a, \infty)) = f^{-1}((-\infty, b)) \cap f^{-1}((a, \infty))$$

which is open by our assumptions. Thus f is continuous.

SOLUTION 4.3.12.

(1) Obviously, in the usual $\mathbb{R}$, the sequence $\left(\frac{1}{n}\right)_{n\geq 1}$ converges to 0.

(2) We have already proved that this topology is not Hausdorff (see Exercise 3.3.22) and hence if this sequence is convergent, it need not have a unique limit.

   The sequence $\left(\frac{1}{n}\right)_{n\geq 1}$ converges to every element of $\mathbb{R}$. To see this, let U be an open set containing x, where $x \in \mathbb{R}$. Hence U^c must be finite. Since

$$\frac{1}{n} \in \mathbb{R} = U \cup U^c,$$

we see that $\frac{1}{n}$ must be in U, for all, but finitely many, $n \in \mathbb{N}$. This proves the convergence of the sequence.

(3) On the contrary of the previous topology, this same sequence does not converge to any point in $\mathbb{R}$. To illustrate this, let us show that $\frac{1}{n}$ does not converge to $a \in \mathbb{R}$ for any a. The question amounts to finding a $U \in \mathcal{V}(a)$ such that

$$\forall N \in \mathbb{N}, \exists n \; (n \geq N \wedge \frac{1}{n} \notin U).$$

It suffices to take $U = \{a\} \in \mathcal{V}(a)$ (which is of course open in this topology) and then

$$\forall N \in \mathbb{N}, \exists n = N \; (n \geq N \wedge \frac{1}{n} \notin \{a\}).$$

   REMARK. In fact, the only convergent sequences in a discrete topological (or metric) space are the constant ones.

(4) In the indiscrete topology, all sequences (and in particular ours) converge to every point in $\mathbb{R}$. For $\mathbb{R}$ is the *only* non-empty open set. Thus it will contain any sequence.

SOLUTION 4.3.13.

(1) Let $x \in \mathbb{R}$. Since $U = (x - 1, x + 1) - K$ is a neighborhood of x and since $\frac{1}{n} \notin U$ for all n, we immediately deduce that $\frac{1}{n} \nrightarrow x$.

   On the contrary, $-\frac{1}{n}$ does converge to 0 in $\mathbb{R}_K$ since any neighborhood of zero will contain infinitely many points of (x_n) by the Archimedean property.

(2) $-K$ is not closed because $-\frac{1}{n} \in -K$ but $-\frac{1}{n} \to 0$, in $\mathbb{R}_K$, and $0 \notin K$.

(3) We leave it to you to show that (x_n) converges to 0.

No, (x_n) cannot have another limit as $\mathbb{R}_\ell$ is Hausdorff (see Exercise 3.3.29).

(4) No, f is not continuous as K is closed in $\mathbb{R}_K$ while its preimage $f^{-1}(K) = -K$ is not closed in $\mathbb{R}_K$.

(5) No, f is not continuous since $[0, \infty)$ is open in $\mathbb{R}_\ell$ and it is not the case for its preimage $f^{-1}([0, \infty)) = (-\infty, 0]$ is not open in $\mathbb{R}_\ell$ (cf. Exercise 3.3.29).

SOLUTION 4.3.14.

(1) Let (x_n) be a convergent sequence to some x. Then

$$\forall U \in \mathcal{V}(x), \ \exists N \in \mathbb{N}, \forall n \ (n \geq N \Rightarrow x_n \in U).$$

In particular, for $U = X \setminus \{x_n : \ x_n \neq x\}$ (which is open and contains x). Thus there exists some N_0 such that for all n :

$$n \geq N_0 \Rightarrow x_n \in X \setminus \{x_n : \ x_n \neq x\}.$$

Therefore, $x_n = x$ for all $n \geq N_0$.

(2) (a) $1 \in [2, 3]'$ since every open set containing 1 intersects $[2, 3]$ (why?).

(b) Since (x_n) takes its values in $[2, 3]$ and since (x_n), if it converges, is eventually a constant, we deduce directly that $x_n \not\to 1$.

(c) The conclusion is: there are sets having a limit point to which no sequence in this set need to converge.

SOLUTION 4.3.15.

(1) A simple application of Exercise 2.5.2 gives, for all n

$$0 \leq |d(x_n, y_n) - d(x, y)| \leq d(x_n, x) + d(y_n, y).$$

Passing to the limit, as n tends to infinity, finishes the proof.

(2) The previous result means that the function d, defined on $X \times X$, is a continuous function (something already known from the metric spaces chapter!).

SOLUTION 4.3.16.

(1) Let us prove this. Assume f is continuous and let $x_n \to x$. We have to prove that $f(x_n) \to f(x)$ in Y. Let U be an open neighborhood of $f(x)$. Then x is in $f^{-1}(V)$ which is open by the continuity of f. Thus, $f^{-1}(U)$ contains all but finitely many terms of (x_n). Accordingly, U contains all but finitely many terms of $f(x_n)$. This means that $f(x_n) \to f(x)$ in Y, establishing the sequential continuity of f.

(2) Let $f : X \to Y$, where X is $\mathbb{R}$ equipped with the co-countable topology and Y is the usual $\mathbb{R}$, be defined by $f(x) = x$. Then the only convergent sequences in X are the eventually constant ones (see Exercise 4.3.14). These sequences also converge in usual $\mathbb{R}$ and hence

$$x_n \to x \text{ in } X \Rightarrow x_n \to x \text{ in } Y \text{ or } f(x_n) \to f(x).$$

This means that f is sequentially continuous. However, it is not continuous for $U = (-1, 1)$ is open in Y and it is not the case for its preimage in X.

(3) Thanks to Question 1, we only prove f is sequentially continuous implies that f is continuous. Assume $f : X \to Y$ (X and Y being two metric spaces) is sequentially continuous and we show that f is continuous and it is better here to use closed sets (why?). Let V be a closed set in Y. We need to establish the closedness of $f^{-1}(V)$ in X. Let $x_n \in f^{-1}(V)$ be converging to $x \in X$. Then $f(x_n)$ is in V for all n and the sequential continuity hypothesis implies that $f(x_n) \to f(x)$. But V is closed and so $f(x) \in V$ or $x \in f^{-1}(V)$. The proof is complete.

SOLUTION 4.3.17. It is known that an open set U in $\mathbb{R}$ is of the form

$$U = \bigcup_{i \in I}(a_i, b_i), \ a_i, b_i \in \mathbb{R}.$$

We have to show that $f(U)$ is open. But since

$$f(U) = f(\bigcup_{i \in I}(a_i, b_i)) = \bigcup_{i \in I} f((a_i, b_i)),$$

we need only show that $f((a_i, b_i))$ are open for every i. WLOG we may assume that f is increasing. Now, since f is *continuous* and *increasing*, we have

$$f((a_i, b_i)) = (f(a_i), f(b_i)), \ \forall i \in I$$

which are all open and hence so is their union. Thus $f(U)$ is open, i.e. f is an open map.

SOLUTION 4.3.18. Let C be a closed set in $\mathbb{R}$. We need to show that $P(C)$ is closed. Let $(y_n) \subset P(C)$ be a converging sequence to y. Hence, there is $(x_n) \subset C$ such that $P(x_n) = y_n$. This implies that $(P(x_n) = y_n)$ is bounded. But since a polynomial has an infinite limit only at $\pm\infty$, we get that (x_n) is also bounded in $\mathbb{R}$. A standard result from the course of calculus (namely the Bolzano-Weierstrass property) tells us that we can extract from (x_n) a convergent subsequence $(x_{n(k)})$.

Call x its limit. Whence, $x \in C$ as C is closed. By the continuity of P (it is a polynomial!), we obtain

$$y \longleftarrow y_{n(k)} = P(x_{n(k)}) \longrightarrow P(x).$$

Since we are in a Hausdorff space, the limit is unique and hence $y = P(x) \in P(C)$, i.e. $P(C)$ is closed which means that P is a closed mapping. The solution is over.

SOLUTION 4.3.19. Call the restriction of f to A, f_A. Let U be an open set in Y. Then $f_A^{-1}(U) = A \cap f^{-1}(U)$ is open in A (in the subspace topology) as $f^{-1}(U)$ is open in X by the continuity of f.

SOLUTION 4.3.20. None of the sets considered in this exercise is closed.

(1) To show that $A = (0,1]$ is not closed in $\mathbb{R}$, it suffices to find a convergent sequence $(x_n)_n$ in A having a limit not belonging to A. Take $x_n = \frac{1}{n}$ which obviously lies in A. Its limit in $\mathbb{R}$ is 0 and it is not in A. Hence A is not closed.

(2) The same arguments (and even the same sequence) apply to show that B is also not closed.

(3) C is not closed. To see that we need a convergent sequence $(x_n, y_n)_n$ in C having a limit outside C. One choice among many is to take

$$(x_n, y_n) = \left(\sqrt{\frac{n}{1+n}}, 0\right) \in C \text{ as } \frac{n}{1+n} + 0 < 1, \ \forall n \in \mathbb{N}.$$

Then, its limit is $(1,0) \notin C$ since $1^2 + 0^2 \not< 1$.

(4) The same method again. The reader may easily show that D is not closed (consider for instance the sequence $\left(\sqrt{1+\frac{1}{n}}, 0\right)_n$).

SOLUTION 4.3.21. First we show that A is not closed. Since $(x_n, y_n) = \left(\frac{n+1}{n}, \frac{n}{n+1}\right) \in A$ $(n \geq 1)$ with limit $(1,1)$ not in A, we easily conclude that A is not closed.

To show that A is not open, we show instead (and equivalently) that $\mathbb{R}^2 \setminus A$ is not closed. Obviously

$$\left(\frac{2n+1}{n}, \frac{n-2}{2n}\right) \notin A, \text{ i.e. } \left(\frac{2n+1}{n}, \frac{n-2}{2n}\right) \in \mathbb{R}^2 \setminus A.$$

However, $\left(\frac{2n+1}{n}, \frac{n-2}{2n}\right) \to (2, \frac{1}{2}) \in A$, i.e. $(2, \frac{1}{2}) \notin \mathbb{R}^2 \setminus A$. Therefore $\mathbb{R}^2 \setminus A$ is not closed, i.e. A is not open.

SOLUTION 4.3.22. A similar idea to the second part of the previous solution may be applied to prove that A is not open. We show that

$\mathbb{R} \setminus A$ is not closed. Consider the sequence $(1 + \frac{1}{p})_{p \geq 1}$. It certainly does not belong to A and hence it belongs to $\mathbb{R} \setminus A$. Its limit in $\mathbb{R}$ is $1 \in A$, i.e. $1 \notin \mathbb{R} \setminus A$. Thus $\mathbb{R} \setminus A$ is not closed.

SOLUTION 4.3.23.

(1) There are different methods to answer this question. We give the following one (the reader may try to give a different proof). Let $(x_n)_n \in A$ such that $x_n \to x$ in X. We need to show that $x \in A$, i.e. $f(x) = g(x)$. Since $(x_n)_n \in A$, $f(x_n) = g(x_n)$. But f and g are both continuous. Hence

$$f(x) \longleftarrow f(x_n) = g(x_n) \longrightarrow g(x).$$

Since Y is Hausdorff (why?), the limit is unique and thus $f(x) = g(x)$, i.e. $x \in A$.

(2) The set B is such that $\overline{B} = X$. Assume that f and g coincide on B and let us show that this forces them to coincide on all of X. Let $x \in X$. Then there exists a sequence $(x_n)_n$ in B such that $x_n \to x$ (in X). Hence one has $f(x_n) = g(x_n)$ and since f and g are continuous,

$$f(x) \longleftarrow f(x_n) = g(x_n) \longrightarrow g(x).$$

Thus f and g coincide everywhere.

REMARK. There is a tempting but completely false proof of the first question. We write A as $(f - g)^{-1}(\{0\})$, then we say that A is closed as it is equal to the inverse image of $\{0\}$ by a continuous function which is $f - g$. There are some false arguments here mainly algebraic ones, e.g. who knows whether 0 is in Y? is " $-$ " defined in Y? (in fact the function $f - g$ may make no sense at all). Of course if $Y = \mathbb{R}$, then this wrong proof becomes a true and nice one.

SOLUTION 4.3.24. If f were continuous at $(0, 0)$, then we would have for any (x_n, y_n) converging to $(0, 0)$, $f(x_n, y_n) \to f(0, 0)$. But $(\frac{1}{n}, \frac{1}{n}) \to (0, 0)$ and $f(\frac{1}{n}, \frac{1}{n}) = \frac{1}{2} \nrightarrow 0$ which means that f is discontinuous at $(0, 0)$ (taking $(\frac{1}{n}, 0) \to (0, 0)$ shows that the limit at $(0, 0)$ does not even exist).

SOLUTION 4.3.25.

(1) Assume that $f : X \to Y$ is a continuous, one-to-one mapping and that Y is Hausdorff. Let us show that X is Hausdorff. Let $x, y \in X$ such that $x \neq y$. Since f is one-to-one, $f(x) \neq f(y)$. But Y is Hausdorff (and $f(x), f(y) \in Y$) and hence

$$\exists U \in \mathcal{V}(f(x)), \ \exists V \in \mathcal{V}(f(y)) \text{ such that } U \cap V = \varnothing.$$

Since U and V are open and f is continuous, $f^{-1}(U)$ and $f^{-1}(V)$ are also open. Since $f(x) \in U$ and $f(y) \in V$, $x \in f^{-1}(U)$ and $y \in f^{-1}(V)$. We also have

$$f^{-1}(U) \cap f^{-1}(V) = f^{-1}(U \cap V) = f^{-1}(\varnothing) = \varnothing.$$

Hence $f^{-1}(U)$ and $f^{-1}(V)$ are two disjoint neighborhoods of the x and y respectively (remember that $x \neq y$). Thus X is Hausdorff.

(2) On $\mathbb{R}$, consider the discrete topology (denoted by Y) and the indiscrete one (denoted X). Now let $f : X \to Y$ be the identity map. Then f is one-to-one but it is *not* continuous (see Exercise 4.3.1). It is also known that Y is Hausdorff whereas X is not.

(3) Keeping the same topologies as in the previous answer but take $f(x) = a$ (the constant function). Then f is continuous but not one-to-one and Y is Hausdorff whilst X is not.

SOLUTION 4.3.26.

(1) The set $Z(g)$ is closed since it is the inverse image of a closed set under a continuous function.

(2) First, we must check that f is well-defined, i.e. its denominator never vanishes. By Exercise 3.3.33 and since A and B are disjoint, we immediately see that $d(x, A)$ and $d(x, B)$ cannot vanish simultaneously and hence

$$\forall x \in X : \ d(x, A) + d(x, B) > 0.$$

Now, since $x \mapsto d(x, A)$ and $x \mapsto d(x, B)$ are (uniformly) continuous by Exercise 2.3.22, the function f, being the quotient of two continuous functions, is continuous.

(3) We have by Exercise 3.3.33

$$f(x) = 0 \Longleftrightarrow d(x, A) = 0$$
$$\Longleftrightarrow x \in \overline{A} = A$$

and hence $f^{-1}(\{0\}) = A$.

A quite similar reasoning applies to show that $f^{-1}(\{1\}) = B$.

(4) We must show that any closed set is the zero set of some continuous function. Let $h : X \to \mathbb{R}$ be a given function. Different cases are to be treated.

(a) If $Z(h) = \varnothing$, take $h(x) = 1$ for all $x \in X$ and this is a continuous function.

(b) If $Z(h) = X$, take $h(x) = 0$ for all $x \in X$ and this is a continuous function.

(c) If $Z(h) \neq X$ (and $Z(h) \neq \varnothing$) is closed, then there exists $a \in X$ such that $h(a) \neq 0$. This implies that the two sets $Z(h)$ and $\{a\}$ are disjoint. Moreover, they are both closed. Define h by

$$h(x) = \frac{d(x, Z(h))}{d(x, \{a\}) + d(x, Z(h))}$$

for all $x \in X$. By Question 2, this is a continuous function. The proof is complete.

(5) Let

$$U = f^{-1}\left(\left(-\frac{1}{5}, \frac{1}{4}\right)\right) \text{ and } V = f^{-1}\left(\left(\frac{1}{3}, \frac{3}{2}\right)\right)$$

Since f is continuous, both U and V are open. They are also disjoint for

$$U \cap V = f^{-1}\left(\left(-\frac{1}{5}, \frac{1}{4}\right)\right) \cap f^{-1}\left(\left(\frac{1}{3}, \frac{3}{2}\right)\right)$$

$$= f^{-1}\left(\left(-\frac{1}{5}, \frac{1}{4}\right) \cap \left(\frac{1}{3}, \frac{3}{2}\right)\right)$$

$$= f^{-1}(\varnothing) = \varnothing.$$

In the end, A is contained in U for $0 \in (-\frac{1}{5}, \frac{1}{4})$ and V contains B as $1 \in (\frac{1}{3}, \frac{3}{2})$ (why?).

SOLUTION 4.3.27. Let $n, m \in \mathbb{N}$ be such that $n \neq m$. It is clear (isn't?) that

$$d_\infty(f_n, f_m) = 1.$$

SOLUTION 4.3.28. We recall that for $f, g \in X$

$$d(f, g) = \int_0^1 |f(x) - g(x)| dx \text{ and } d'(f, g) = \sup_{x \in [0,1]} |f(x) - g(x)|.$$

First, we prove the closedness of A with respect to d'. Let $f \in \overline{A}$. Then for some $f_n \in A$, f_n converges uniformly to f. This implies two things. First, that f must be continuous (a well-known result from the course of advanced calculus, or from Chapter 8).

Second that

$$\forall x \in [0, 1] : \lim_{n \to \infty} f_n(x) = f(x).$$

But, since $f_n(0) = 0$, we have $f(0) = 0$ too. Thus $f \in A$.

Now we show that A is dense in X with respect to d. We need only show that $X \subset \overline{A}$. Let $f \in X$, i.e. f is continuous. Consider the sequence of functions f_n defined by

$$f_n(x) = \begin{cases} xe^n f(e^{-n}), & 0 \le x \le e^{-n}, \\ f(x), & e^{-n} \le x \le 1. \end{cases}$$

The continuity of f implies that of the f_n. Also $f_n(0) = 0$. Hence $f_n \in A$ for all $n \in \mathbb{N}$. It only remains to show that $d(f_n, f) \to 0$ as n tends to infinity. Let $n \in \mathbb{N}$. We have

$$d(f_n, f) = \int_0^{e^{-n}} |f_n(x) - f(x)| dx + \int_{e^{-n}}^1 |f_n(x) - f(x)| dx$$

$$= \int_0^{e^{-n}} |xe^n f(e^{-n}) - f(x)| dx.$$

But for all $(x, n) \in [0, e^{-n}] \times \mathbb{N}$ we have

$$|xe^n f(e^{-n}) - f(x)| \le |xe^n f(e^{-n})| + |f(x)| \le e^{-n} e^n |f(e^{-n})| + |f(x)|$$

$$\le 2 \sup_{x \in [0, e^{-n}]} |f(x)| \le 2 \sup_{x \in [0,1]} |f(x)|.$$

Thus

$$d(f_n, f) \le 2 \sup_{x \in [0,1]} |f(x)| \int_0^{e^{-n}} dx = 2e^{-n} \sup_{x \in [0,1]} |f(x)| \to 0 \text{ as } n \to \infty.$$

Therefore, $f \in \overline{A}$.

SOLUTION 4.3.29.

(1) We can prove A_a is closed as done in the foregoing exercise. Alternatively, we can do the following. Let d be the supremum metric on X. Let $f_n \in A_a$ such that $d(f_n, f) \to 0$ as n tends to infinity. Then f is continuous on $[0, 1]$. We also have

$$|f(a)| \le |f_n(a) - f(a)| + |f_n(a)| = |f_n(a) - f(a)| \le d(f_n(a), f(a)) \to 0$$

as n tends to infinity. This gives $f \in A_a$.

(2) Observe that for each $a \in I$, B reduces to some A_a. Thus B may be written as the arbitrary intersection of closed sets of the form A_a, i.e.

$$B = \bigcap_{a \in I} A_a \text{ is closed.}$$

SOLUTION 4.3.30.

(1) First, we show the given projections are continuous. We show that p is continuous . Let U be an open set in X. We need to show that $p^{-1}(U)$ is open in $X \times Y$. We have

$$p^{-1}(U) = \{(x,y) \in X \times Y : p(x,y) = x \in U\} = U \times Y$$

which is open in $X \times Y$. Thus p is continuous. The proof of the continuity of q is very akin to that of p.

Second, we note that the set A (given in the hint) is closed (see the next coming exercise for the proof that A is closed). Now we have

$$p(A) = \left\{p\left(x, \frac{1}{x}\right) : x \neq 0\right\} = \mathbb{R}^*.$$

Thus $p(A)$ is not closed and hence p is not closed. The same arguments apply for q.

(2) Let U and V be open sets in X and Y respectively. We have

$$p(U \times V) = \{p(x,y) = x : (x,y) \in U \times V\} = U.$$

Now, any open set Ω in $X \times Y$ is of the form $\cup_{i \in I}(U_i \times V_i)$ where U_i and V_i are open sets in X and Y respectively. Moreover,

$$p(\Omega) = p\left(\bigcup_{i \in I}(U_i \times V_i)\right) = \underbrace{\bigcup_{i \in I} p(U_i \times V_i)}_{\text{open in } X}$$

and hence p is an open mapping. The proof of the openness of q is very similar to that of p.

SOLUTION 4.3.31.

(1) We are required to prove that X is Hausdorff if and only if $\triangle$ is closed, that is, if and only if $\triangle^c$ is open. Assume X is separated. Let $(x,y) \in \triangle^c$. Then $x \neq y$. But X is Hausdorff and hence

$$\exists (U,V) \in \mathcal{V}(x) \times \mathcal{V}(y) : U \cap V = \varnothing.$$

Now, $U \times V$ is open in $X \times X$, it contains (x,y) and $(U \times V) \cap \triangle = \varnothing$ which implies that $U \times V \subset \triangle^c$. By Test 7, we get that $\triangle^c$ is open or that $\triangle$ is closed.

Conversely, suppose that $\triangle$ is closed. To show that X is separated, let $x \neq y$. Then $(x,y) \in \triangle^c$. But $\triangle^c$ is a union of basis elements and hence

$$\exists U, V \in X \times X : (x,y) \in U \times V \subset \triangle^c.$$

This implies that $U \cap V = \varnothing$ (if a were in both U and V, then $(a, a) \in U \times V$ but it would not be in $\triangle^c$, a contradiction!). Since U and V are open sets that contain x and y respectively, we immediately see that X Hausdorff, establishing the result.

(2) The function $(f, g) : X \to Y \times Y$ is continuous (this will be illustrated below). Now, let $\triangle$ be the diagonal of Y. Then

$$[(f, g)]^{-1}(\triangle) = \{x \in X : f(x) = g(x)\}$$

is closed, being the preimage of a closed set (which one and why?) by a continuous function.

SOLUTION 4.3.32.

(1) (a) The "if" part. Let $x \in A$. Adopting the notations of Exercise 4.3.30 we can write $g(x) = p(f(x))$ and $h(x) = q(f(x))$. Since f, p and q are continuous, so are g and h.

(b) The "only if" part. Assume that h and g are both continuous. Let U be an open set in X let V be an open set in Y. We first show that $f^{-1}(U \times V)$ is open in A. We have

$$\begin{aligned}
x \in f^{-1}(U \times V) &\Leftrightarrow f(x) \in U \times V \\
&\Leftrightarrow g(x) \in U \wedge h(x) \in V \\
&\Leftrightarrow x \in g^{-1}(U) \wedge x \in h^{-1}(V) \\
&\Leftrightarrow x \in g^{-1}(U) \cap h^{-1}(V).
\end{aligned}$$

This shows that $f^{-1}(U \times V) = g^{-1}(U) \cap h^{-1}(V)$. Since g and h are continuous, both $g^{-1}(U)$ and $h^{-1}(V)$ are open in A. Hence $f^{-1}(U \times V)$ is also open in A. However, every open set in $X \times Y$ is a union of sets of the form $U \times V$ where U is open in X and V is open in Y. If Ω is open in $X \times Y$, then $\Omega = \bigcup_{i \in I} U_i \times V_i$ and

$$f^{-1}(\Omega) = f^{-1} \left(\bigcup_{i \in I} U_i \times V_i \right) = \underbrace{\bigcup_{i \in I} f^{-1}(U_i \times V_i)}_{\text{open in } A}.$$

Thus f is continuous.

(2) The answer is no in this case, i.e. a function may well have partial continuous functions without being continuous itself. The following example elucidates that. Let $f : \mathbb{R}^2 \to \mathbb{R}$ be

defined by

$$f(x,y) = \begin{cases} \frac{2xy}{x^2+y^2}, & (x,y) \neq (0,0), \\ 0, & (x,y) = (0,0). \end{cases}$$

The reader can easily check that $x \mapsto f(x,y)$ and $y \mapsto f(x,y)$ are both continuous on $\mathbb{R}$ while f is not continuous at $(0,0)$.

REMARK. The result in Question 1 can be generalized to an infinite cartesian product $\prod_{i \in I} X_i$ where the X_is are topological spaces if it is given the product topology (see **[10]**). If, however, we endow it with the box topology (for which the basis elements are of the form $\prod_{i \in I} U_i$ where the U_is are open in the X_is), then the result in Question 1 fails to hold.

As a counterexample (borrowed from **[10]**), take $f : \mathbb{R} \to \mathbb{R}^\omega$, where $\mathbb{R}^\omega = \mathbb{R} \times \mathbb{R} \times \cdots \times \mathbb{R} \times \cdots$ (a countably infinite product), defined by

$$f(x) = (x, x, \cdots, x, \cdots).$$

By the generalization alluded to just above f is continuous with respect to the product topology since all its components are continuous. However, f is not continuous with respect to the box topology since

$$\Omega = (-1,1) \times \left(-\frac{1}{2}, \frac{1}{2}\right) \times \left(-\frac{1}{3}, \frac{1}{3}\right) \times \cdots \times \left(-\frac{1}{n}, \frac{1}{n}\right) \times \cdots$$

is open in $\mathbb{R}^\omega$ while its preimage $f^{-1}(\Omega)$ is not open in $\mathbb{R}$. For if it were and since $(0, 0, \cdots, 0 \cdots) \in \Omega$, we would have for some $r > 0$,

$$(-r, r) \subset f^{-1}(\Omega) \implies f(-r, r) \subset \Omega.$$

Applying the nth projection to the previous inclusion would yield

$$(-r, r) \subset \left(-\frac{1}{n}, \frac{1}{n}\right), \ \forall n \in \mathbb{N},$$

which contradicts the Archimedes axiom.

In this book we shall not use topologies of infinite product of topological spaces. The reader is referred to **[10]** for more on the infinite product topology.

SOLUTION 4.3.33.

(1) The proof is based on Exercise 4.3.31. We know that $p : (x,y) \mapsto x$ and $q : (x,y) \mapsto y$ are both continuous. Then $f \circ p : (x,y) \mapsto f(x)$ is continuous too. Thus by Exercise 4.3.31, the set

$$\{(x,y) \in X \times Y : (f \circ p)(x,y) = q(x,y)\},$$

which is nothing but the graph of f, is closed. This finishes the proof.

(2) The answer is yes. In usual $\mathbb{R}$, consider
$$f(x) = \begin{cases} 0, & x = 0 \\ \frac{1}{x}, & x \in \mathbb{R}^*. \end{cases}$$

Then f is defined on $\mathbb{R}$ and it takes its value in $\mathbb{R}$ which is Hausdorff. Obviously f is not continuous at $x = 0$. Its graph, given by
$$G_f = \{(x, f(x)) : x \in \mathbb{R}\} = \{(0,0)\} \cup \left\{ \left(x, \frac{1}{x}\right) : x \in \mathbb{R}^* \right\},$$

is closed. To show this, we observe that since $\{(0,0)\}$ is closed in $\mathbb{R}^2$, we need only show that $B = \left\{ \left(x, \frac{1}{x}\right) : x \in \mathbb{R}^* \right\}$ is closed. But since $xy = 1 \Rightarrow x \neq 0$, we can write
$$B = \{(x,y) \in \mathbb{R}^2 : xy = 1\}$$

and this is easily seen to be closed (why?). Finally, remember that $\mathbb{R}$ is Hausdorff.

4.4. Hints/Answers to Tests

SOLUTION 23. No! f is not bijective...

SOLUTION 24. It is closed since $A = \varnothing$ (the empty set) and this is the only reason why it is closed!...

SOLUTION 25. The answer is no. Call the co-finite topology X and the other one Y. Then $id : X \to Y$ is not continuous for $\{a\}$ is open in Y but it is not open in X. Similar arguments show that $id : Y \to X$ is not continuous either...

SOLUTION 26. The given sequence does not converge to any point x in this topology since
$$\exists U = \{x, a\} \in \mathcal{V}(x) : \forall N \in \mathbb{N}, \exists n \ (n \geq N \wedge \frac{1}{n} \notin U)...$$

SOLUTION 27. No! $[0, 1)$ is an open set that contains 0...

SOLUTION 28. The non-empty open sets in Y are $\{1\}$ and Y. So if U is one of the latter open sets, then $f^{-1}(U) = A$ or X. What is left to do should be clear to the reader...

SOLUTION 29. f is continuous iff it is constant...

SOLUTION 30.

(1) Should be a routine by now...

(2) No, it is not...

SOLUTION 31. The projection $p : X \times X \to X$ is a homeomorphism (is it not?)...

SOLUTION 32. Yes. Do it!...

SOLUTION 33. Well, do it...

SOLUTION 34. Very easy!...

CHAPTER 5

Compact Spaces

5.2. True or False: Answers

ANSWERS.

(1) True! If we want to show that $\{U_i\}_{i \in I}$ covers the whole space X, then we may write $X = \cup_{i \in I} U_i$ for we always have $X \supset \cup_{i \in I} U_i$.

(2) The answer is negative. In the definition of a compact set, it asked to verify that **every** open cover has a finite subcover.

(3) Using closed covers in the definition of compact spaces would be of little interest. For instance, in Hausdorff spaces, which is already a large class of interesting topological spaces, the only compact spaces (using closed covers) would be the finite ones. For if X is a Hausdorff space, then $\{\{x\}\}_{x \in X}$ is a closed cover for X and we immediately see that X is compact if it is finite.

(4) True if X is compact. In fact, we have an equivalence.

The answer is, however, not true in general. For instance, in $\mathbb{R}$ (which is not compact with respect to the usual topology), let $A_n = [n, \infty)$. Then, for each $n \in \mathbb{N}$, A_n are non-empty, closed in $\mathbb{R}$, $A_{n+1} \subset A_n$ but

$$\bigcap_{n \in \mathbb{N}} A_n = [1, \infty) \bigcap \cdots \bigcap [n, \infty) \bigcap \cdots = \varnothing.$$

(5) The answer is no! In the induced usual metric of $X = [1, \infty)$, let $A_n = [n, \infty)$. Let $f : X \to X$ defined for all $x \geq 1$ by $f(x) = 1$. Then f is continuous. Besides, the sequence (A_n) is non-empty and decreasing for all n. But

$$\bigcap_{n \in \mathbb{N}} A_n = \varnothing \text{ and so } f(\varnothing) = \varnothing$$

whereas $f(A_n) = \{1\}$ and so

$$\varnothing \neq \bigcap_{n \in \mathbb{N}} f(A_n) = \{1\}.$$

The result is, however, true if X is assumed to be compact. See Exercise 5.3.23.

(6) The answer is yes. Take any open cover in any topological space. Then ∅ will always be contained in any finite union of elements of that cover.

(7) In the usual topology of $\mathbb{R}$ (the Heine-Borel theorem). But as soon as we leave the usual topology, this result may fail to hold. For instance, in the discrete topology, $[a, b]$ is not compact any more (see Exercise 5.3.2).

(8) False! Consider the topology of Exercise 3.5.7. Since $\{1\}$ is finite, it is compact. However, $\overline{\{1\}} = \mathbb{N}$ is not compact. For a proof see Test 38.

(9) The answer is no. In the usual topology, consider $f : \mathbb{R} \to \mathbb{R}$ defined by $f(x) = 1$ for all $x \in \mathbb{R}$. Then $A = [0, 1]$ is compact in $\mathbb{R}$ but $f^{-1}(A) = \mathbb{R}$ is not compact.

In Exercise 5.5.10, some condition implying the compactness of $f^{-1}(A)$ is given.

(10) We try to give an exhaustive comment on this question. The well known-result says that a set in $\mathbb{R}$ equipped with the usual topology (or $\mathbb{R}^n$ equipped with the euclidian metric) is closed if and only if it is bounded and closed. This result need not hold if we change $\mathbb{R}$ by $\mathbb{Q}$ even if we give $\mathbb{Q}$ the induced usual topology (see Exercise 5.3.17 where a closed and bounded set is not compact). Even in $\mathbb{R}$ equipped with a metric different from the usual metric, a closed and bounded set does not have to be compact (see also Exercise 5.3.18). Also, in an infinite dimensional space, a closed and bounded set need not be compact either.

It should also be remembered that in any metric space every compact subspace is closed and bounded.

(11) The answer is yes. The reason is simple. Every open cover for X with respect to T is one for X with respect to T'.

The converse is not true. The usual topology is finer than the co-finite topology (on $\mathbb{R}$). However, $\mathbb{R}$ is compact in the co-finite topology and it is not in the usual topology.

(12) The problem with the reasoning is purely algebraic. More precisely,

$$\mathbb{R} \not\subset \bigcup_{i=1}^{n} U_i \not\Rightarrow \mathbb{R} \subset \left(\bigcup_{i=1}^{n} U_i \right)^c$$

(remember that if $A \not\subset B$ and A and B are *disjoint*, then $A \subset B^c$).

For a correct proof of the compactness of $\mathbb{R}$ in the co-finite topology, see Exercise 5.3.11.

(13) First, $\mathbb{R}$ has to be equipped with a metric to be able to talk about the possible boundedness of f. Second, what is the topology given to $[a, b]$? In general, the answer is no as showed by the following example: let $f : [0, 1] \to \mathbb{R}$ defined by

$$f(x) = \left\{ \begin{array}{ll} \frac{1}{x}, & x \in (0, 1], \\ 1, & x = 0. \end{array} \right.$$

If we equip $[0, 1]$ with the discrete topology (and $\mathbb{R}$ with the usual metric), then f will be continuous and f is clearly unbounded. What went wrong with our example is the fact that $[0, 1]$ is *not compact* in the discrete topology (see Exercise 5.3.2).

(14) If considered as a subspace of usual $\mathbb{R}$, then $(0, 2)$ is relatively compact. But, if considered, for instance, as a subspace of itself, then it is clearly not relatively compact.

(15) Not always! For a counterexample, see Exercise 5.3.20. We note that in *normed vector spaces*, the closed unit ball is compact iff the space in questions has a finite (algebraic) dimension. It may be proved in an introductory functional analysis course that a normed vector space is locally compact iff the closed unit ball is compact.

(16) The answer is no! For instance, in $\mathbb{R}$ endowed with the co-finite topology, every subset is compact (see the remark below the solution of Exercise 5.3.11). Hence $\mathbb{R}^+$ is compact but it is not closed. Also observe that this topology is not Hausdorff.

(17) True. Let us show that. Let f be a homeomorphism between two topological spaces X and Y. Assume that X is compact. Since f is continuous, $f(X)$ is compact. Since f is onto $Y = f(X)$. Thus Y is compact. The same idea of proof can be applied to f^{-1} (in lieu of f) so that if Y is compact, so is X.

(18) True. Let us prove it. Let f be a homeomorphism between two topological spaces X and Y. Assume that X is sequentially compact. Let (y_n) be a sequence in Y. Since f is bijective, $f^{-1}(y_n)$ is a well-defined sequence (into X) which we denote by x_n. Since X is sequentially compact, $(x_n = f^{-1}(y_n))$ contains a subsequence, which we denote by $x_{n(k)}$, that converges to some x in X. Since f is continuous, we immediately see that

$$f(x_{n(k)}) = f(f^{-1}(y_{n(k)})) = y_{n(k)} \longrightarrow f(x).$$

Therefore, (y_n) contains a convergent subsequence. Applying the previous proof to f^{-1} (instead of f), we get that whenever Y is sequentially compact, then so is X.

(19) False! For a counterexample the reader is referred to Exercise 5.3.25.

(20) True! The reader is asked to give a proof in Exercise 5.5.15.

(21) False! A counterexample will be encountered in Exercise 5.3.26.

(22) True. Let us show that. Let $A = \{x_1, x_2, \cdots, x_n\}$ be this finite part. Then obviously

$$A \subset \bigcup_{1 \leq i \leq n} \{x_i\} \text{ and } d(\{x_i\}) = 0, \ \forall i = 1, 2, \cdots, n.$$

In fact, a stronger result (in a restrained context) holds: *Every bounded set in $\mathbb{R}^n$ is totally bounded* (the reader is requested to give a proof in Exercise 5.5.18).

If the metric space is arbitrary, then this may not be true (see Exercise 7.5.13). Rememebrer that the converse is always true, that is, *any totally bounded set is bounded.*

(23) **False!** Total boundedness is not a topological property. This is its main weakest point. As a counterexample, consider the function $f : (0, 1) \to (1, +\infty)$ defined by $f(x) = \frac{1}{x}$. Then f is a homeomorphism, $(0, 1)$ is totally bounded whereas $(1, +\infty)$ is not for it is not bounded.

5.3. Solutions to Exercises

SOLUTION 5.3.1. Let $\mathcal{U} = \{(-n, n)\}_{n \in \mathbb{N}}$ be an open cover of $\mathbb{R}$ (cf. Exercise 1.2.6). Now no finite subcollection of $\mathcal{U}$ can cover all of $\mathbb{R}$. For if $\mathcal{U}_p = \{(-n_i, n_i)\}_{1 \leq i \leq p}$ is a finite subcover, then its union will be of the form $(-N, N)$ where $N = \max_{1 \leq i \leq p} n_i$.

For $[0, +\infty)$, we may consider the open cover $\mathcal{U} = \{(-1, n)\}_n$. It is in effect a cover since $[0, +\infty) \subset \bigcup_{n=1}^{\infty} (-1, n)$. Now any finite subcollection of $\mathcal{U}$ is of the form $\{(-1, n_1), (-1, n_2), \cdots, (-1, n_p)\}$ and its union is $(-1, N)$, $N = \max_{1 \leq i \leq p} n_i$. Lastly, it is plain that

$$[0, +\infty) \not\subset (-1, N), \ \forall N \in \mathbb{N}.$$

As for $(0, 1)$, the reader may take $\mathcal{U} = \left(0, 1 - \frac{1}{n}\right)_{n \geq 2}$ (or just $\left(\frac{1}{n}, 1\right)_{n \geq 2}$) and then show that $(0, 1)$ is not compact.

SOLUTION 5.3.2.

(1) $\mathbb{Q}$ is not compact as it is not closed in $\mathbb{R}$ (or since it is not bounded).

(2) $A = \{\frac{1}{n} : n \in \mathbb{N}\}$ is not compact since it is not closed.

(3) $\mathbb{Q} \cap [0,1]$ is not closed in $\mathbb{R}$ and hence it is not compact.

(4) In fact $[a,b]$ is not compact in the discrete topology and neither is $\mathbb{R}$ nor is an infinite set. We only do that in the case of $[a,b]$. In $(\mathbb{R}, \mathcal{P}(\mathbb{R}))$ the singleton $\{x\}$ is open and hence $\mathcal{U} = \{x\}_{x \in [a,b]}$ is an open cover of $[a,b]$ and every finite subcollection of it will never cover $[a,b]$. Therefore, $[a,b]$ is not compact in $(\mathbb{R}, \mathcal{P}(\mathbb{R}))$.

(5) Since we are now in the standard topology of $\mathbb{R}^2$, then to show the compactness of the given sets it suffices to show that they are both closed and bounded in $\mathbb{R}^2$ and this in all the remaining questions of this exercise. The set A is obviously closed and bounded and hence it is compact in $\mathbb{R}^2$. As for B and C none of them is compact since they are not closed (cf. Exercise 4.3.20) and C is even unbounded .

(6) A is not compact as it is not bounded in $\mathbb{R}^2$ (it cannot be contained in a ball in $\mathbb{R}^2$ of finite radius, show it!).

(7) Both A and B are not compact since A is closed and not bounded while B is bounded but not closed. Let us show that. The set A can be modified to be written as

$$A = \{(x,y) \in \mathbb{R}^2 : x \geq 0, xy = 1\}$$

since $xy = 1$ implies $x \neq 0$. Now we can easily show that A is closed. However it is not bounded as it cannot be contained in a ball of finite radius. The set B is obviously bounded since $B \subset (0,1] \times [-1,1]$. Let us show that it is not closed. The sequence $(\frac{1}{n\pi}, 0)$ does belong to B while its limit (in $\mathbb{R}^2$) is $(0,0)$ is not in B.

(8) The reader may show that A is actually equal to $B_c(0_{\mathbb{R}^2}, 1)$ (the closed ball in $\mathbb{R}^2$ of center $(0,0)$ and of radius 1). Thus A is compact.

SOLUTION 5.3.3.

(1) We use the Euclidean metric (we could have used the sup metric or the taxi cab metric too for they are equivalent on $\mathbb{R}^n$. See Exercise 2.3.9). The set A is closed (why?) but it is not bounded. Let us show that. The set A cannot be bounded since for all $M > 0$,

$$(M, \sqrt[3]{1-M}) \in A \text{ and } d_2((M, \sqrt[3]{1-M}), (0,0))^2 \geq M^2.$$

Therefore, A is not compact in usual $\mathbb{R}^2$.

(2) Now we use the taxicab metric. The set B is compact in $\mathbb{R}^2$. It is obviously closed (is it not?). Since $x^4 + y^2 = 1$, $|x| \leq 1$ and $|y| \leq 1$ (why?). Whence

$$\forall x, y \in A: \ d_1((x,y),(0,0)) = |x| + |y| \leq 2,$$

proving the boundedness of B and hence its compactness.

SOLUTION 5.3.4. Set

$$U_n = \left\{ \begin{array}{ll} \left(\frac{2}{3}, \frac{3}{2}\right), & n = 1, \\ \left(\frac{1}{n+1}, \frac{1}{n-1}\right), & n \geq 2. \end{array} \right.$$

Observe that for any $n \geq 1$, U_n is open. It is also clear that

$$A = \left\{ \frac{1}{n} : \ n \in \mathbb{N} \right\} \subset \bigcup_{n \geq 1} U_n.$$

Hence $\mathcal{U} = \{U_n\}_n$ is an open cover for A. Since each U_n contains only one point of A and since A is *infinite*, we conclude that no finite subcover of $\mathcal{U}$ can contain A. Thus A is not complete.

SOLUTION 5.3.5. Two methods are given (another one will follow in Exercise 7.3.23).

(1) *A method based on the completeness axiom:* Let $\mathcal{U} = \{U_i\}_{i \in I}$ be an open cover of $[a, b]$. Set

$A = \{x \in [a, b] : \ [a, x]$ can be covered by a *finite* subcollection of $U_i\}$.

It is clear that me have to show that $b \in A$. We first note that A is non void for $a \in A$ because $[a, a] = \{a\}$ is covered by one U_i for some $i \in I$. Besides, A is bounded above by b as $A \subset [a, b]$. Thus, A has the least upper bound property. Hence its sup, denoted by M, satisfies $M \leq b$. If we come to show that $M \in A$ and $M = b$, then we are done, i.e. $[a, b]$ will be compact.

Let us first show that $M \in A$. Since $M \in [a, b]$, there exists some $j \in I$ such that $M \in U_j$. But U_j is open in the usual topology of $\mathbb{R}$ and so

$$\exists r > 0: \ (M - r, M + r) \subset U_j.$$

Moreover, by definition of the least upper bound we have

$$\exists x \in A: \ M - r < x \leq M.$$

Since $x \in A$, $[a, x]$ is covered by a finite number of U_i ($i = 1, 2, \cdots, n$, say), $\{U_1, U_2, \cdots, U_n, U_j\}$ covers the interval $[a, M]$

for

$$[a, M] = [a, x] \cup [x, M] \subset [a, x] \cup (M - r, M + r) \subset \underbrace{\bigcup_{i=1}^{n} U_i \cup U_j}_{\text{a finite union!}}.$$

Therefore, $M \in A$. To finish the proof, we need to verify that $M = b$. Assume $M \neq b$, i.e. $M < b$. Pick a y such that $y < b$ and $M < y < M + r$. Then we immediately see that $\{U_1, U_2, \cdots, U_n, U_j\}$ covers $[a, y]$ too (why?), i.e. $y \in A$. We have then reached a contradiction as we have $y \in A$ and $M < y$!

Thus $b = M \in A$, i.e. $[a, b]$ is covered by a *finite* subcollection of U_i, that is, $[a, b]$ is compact. The proof is complete.

(2) *A method based on the nested interval property*: Assume $[a, b]$ is not compact, i.e. there is an open cover $U = \{U_i\}_{i \in I}$ from which no finite subcover can be extracted. We bisect $[a, b]$ at its middle point, i.e. at $\frac{1}{2}(a + b)$. We then obtain the two closed intervals

$$\left[a, \frac{a+b}{2}\right] \text{ and } \left[\frac{a+b}{2}, b\right]$$

(observe that their length is $(b - a)/2$).

One (at least!) of the two segments cannot be covered by a finite number of U_i otherwise the two intervals and hence their union will be covered by a finite number of U_i! Call this interval I_1 and write it as $[a_1, b_1]$. Now, bisect it into two intervals as done previously and obtain the two intervals

$$\left[a_1, \frac{a_1 + b_1}{2}\right] \text{ and } \left[\frac{a_1 + b_1}{2}, b_1\right]$$

(with this time intervals having for length $(b - a)/4$ regardless of what a_1 and b_1 can be). As before, at least one of them cannot be covered by a finite number of U_i. Continuing this process, we obtain a sequence of closed and bounded intervals (I_n) (with $I_0 = [a, b]$) verifying

(a) (I_n) is decreasing;
(b) the length of I_n is $(b - a)/2^n$ for each n;
(c) each I_n cannot be covered by a finite number of U_i.

Now, by construction, the sequences (a_n) and (b_n) satisfy

$$a \leq a_n \leq a_{n+1} \leq b_{n+1} \leq b_n \leq b$$

for all n. Thus (a_n) is increasing and bounded above by b and (b_n) is decreasing and bounded below by a. Hence they both converge to different limits, but the condition $b_n - a_n = (b - a)/2^n$ ensures that this limit must be the same, call it l. It is certainly (by the sandwich rule!) inside $[a, b]$ and so it lies in some U_j, $j \in I$, of the initial open cover. By the openness of U_j we have

$$\exists r > 0 : \; (l - r, l + r) \subset U_j.$$

Since $(b - a)/2^n$ tends to zero as n goes to infinity, we have for large n, $(b - a)/2^n < r$ which gives

$$[a_n, b_n] \subset (l - r, l + r)$$

(remember that $l \in [a_n, b_n]$ for all n). Therefore, we have reached the desired contradiction for $[a_n, b_n]$ cannot be covered by a *finite* number of U_i. The proof is complete.

SOLUTION 5.3.6. We answer both questions directly. We claim that any set containing K is not compact. The proof is simple. Let A be a set which contains K. Since $K' = \varnothing$ (see Exercise 3.3.30), we immediately deduce that A is not limit point compact and hence it is not compact.

SOLUTION 5.3.7. We write what $x_n \to a$ means in a topological space

$$x_n \longrightarrow a \iff \forall U \in \mathcal{V}(a), \exists N \in \mathbb{N}, \forall n \; (n \geq N \Rightarrow x_n \in U).$$

Now let $\mathcal{U} = \{U_i\}_{i \in I}$ be an open cover of $A = \{x_n : \; n \in \mathbb{N}\} \cup \{a\}$. Hence $A \subset \bigcup_{i \in I} U_i$. We can write the following

$$A = \{x_n : \; n < N\} \cup \{x_n : \; n \geq N\} \cup \{a\}.$$

With the x_n for $n < N$ we can associate $N - 1$ elements of the cover $\mathcal{U}$. Also we observe that $\{a\}$ is contained in some U_j and since the latter is open (and contains a) then it is a neighborhood of a and hence by definition of the limit (see above) we also have $x_n \in U_j$ for $n \geq N$. Therefore we deduce that $A \subset \bigcup_{i=1}^{N-1} U_i \cup U_j$, establishing the compactness of A.

SOLUTION 5.3.8.

(1) Let A be a finite set having n elements. Hence we may write

$$A = \{x_1, x_2, \cdots, x_n\}$$

where all x_i $(1 \leq i \leq n)$ belong to X. Let $\mathcal{U} = \{U_i\}_{i \in I}$ be an open cover of A, i.e.

$$A = \{x_1, x_2, \cdots, x_n\} \subset \bigcup_{i \in I} U_i.$$

Hence, each x_i will belong to some U_i and thus

$$A = \{x_1, x_2, \cdots, x_n\} \subset U_1 \cup U_2 \cup \cdots \cup U_n,$$

establishing the compactness of A.

(2) Question 4 of Exercise 5.3.2 combined with the preceding question does the job.

SOLUTION 5.3.9.

(1) Let A and B be compact. Let $\mathcal{U} = \{U_i\}_{i \in I}$ be an open cover of $A \cup B$, i.e.

$$A \cup B \subset \bigcup_{i \in I} U_i.$$

Hence

$$A \subset \bigcup_{i \in I} U_i \text{ and } B \subset \bigcup_{i \in I} U_i.$$

This tells us that $\mathcal{U} = \{U_i\}_{i \in I}$ is an open cover for both A and B. Since they are compact, we have

$$A \subset \bigcup_{i=1}^{n} U_i \text{ and } B \subset \bigcup_{j=1}^{m} U_j$$

which yield

$$A \cup B \subset \underbrace{\bigcup_{i=1}^{n} U_i \cup \bigcup_{j=1}^{m} U_j}_{\text{a finite union}}.$$

Therefore, $A \cup B$ is compact.

The previous result is not true for an arbitrary union. For instance, in standard $\mathbb{R}$, the intervals $[-n, n]$ are all compact for any $n \in \mathbb{N}$. However, $\bigcup_{n=1}^{\infty} [-n, n] = \mathbb{R}$ is not compact.

(2) Let $(A_i)_{i \in I}$ be an arbitrary family of compact sets in a Hausdorff space. Since they are all compact, they are all closed.

Hence $\bigcap_{i \in I} A_i$ is closed. But for some (and all) $j \in I$ one has

$$\underbrace{\bigcap_{i \in I} A_i}_{\text{closed}} \subset \underbrace{A_j}_{\text{compact}} .$$

Thus $\bigcap_{i \in I} A_i$ is compact since a closed subset of a compact set is compact.

SOLUTION 5.3.10. Let $t \in X$. Recall that $x \mapsto d(x,t)$ is continuous on X (see Exercise 2.3.22). Since A is compact, $\inf_{x \in A} d(x,t)$ is attained so that

$$\exists x_0 \in A : \inf_{x \in A} d(x,t) = d(x_0, t),$$

which, in its turn, implies that:

$$\bigcup_{x \in A} B_c(x,r) = \{t \in X : \inf_{t \in A} d(x,t) \le r\}.$$

The latter set is closed for the function $t \mapsto \inf_{x \in A} d(x,t)$ is continuous on $\mathbb{R}$ (is it not?) and $(-\infty, r]$ is closed.

SOLUTION 5.3.11. Yes, $\mathbb{R}$ is compact in the co-finite topology. Let $\mathcal{U} = \{U_i\}_{i \in I}$ be an open cover of $\mathbb{R}$ in X, i.e. $\mathbb{R} = \bigcup_{i \in I} U_i$. We need to show that $\mathbb{R}$ can be covered by a finite subcollection of $\mathcal{U}$. We have $\mathbb{R} = U_j \cup U_j^c$ $(j \in I)$ and since U_j^c is finite, it has the form $\{x_1, x_2, \cdots, x_p\}$, say, where all x_i are real. Hence

$$\{x_1, x_2, \cdots, x_p\} \subset U_1 \cup U_2 \cup \cdots \cup U_p$$

and thus

$$\mathbb{R} = U_j \cup U_j^c \subset U_j \cup U_1 \cup U_2 \cup \cdots \cup U_p.$$

This shows that $\mathbb{R}$ is compact in the co-finite topology.

REMARK. The same method applies to show that any infinite set (finite sets are already compact!) is compact in the co-finite topology.

SOLUTION 5.3.12.

(1) $\mathbb{R}$ is not compact. Let $U = \{x_n : n \in \mathbb{N}\}$ be a countable set. Then set

$$U_n = U^c \cup \{x_1, x_2, \cdots, x_n\}.$$

Hence

$$\bigcup_{n=1}^{\infty} U_n = U^c \cup U = \mathbb{R},$$

i.e. $\{U_n\}_n$ constitutes an *open* cover for $\mathbb{R}$. However, no finite subcollection of $\{U_n\}_n$ can cover all of $\mathbb{R}$. To see this explicitly,

$$\bigcup_{p=1}^{N} U_{n(p)} \neq \mathbb{R} \text{ as, for instance, } x_{N+1} \notin \bigcup_{n=1}^{N} U_{n(p)}$$

where $N = \max_{1 \leq p \leq N} n(p)$.

(2) No, for the same reason as before. In fact, the previous method can be applied to show that any infinite set is not compact in the co-countable topology.

(3) No and also as before. An alternative way of seeing this is the following: If A is countable, then the induced topology is the discrete one and in a discrete topology a set is compact iff it is finite!

SOLUTION 5.3.13.

(1) Let us show that T is a topological space on $\mathbb{R}$. $\varnothing$ belongs to T by definition of T and $\mathbb{R} \in T$ since $\mathbb{R}^c = \varnothing$ is compact. Let U and V be both in T.

    (a) If U or V is empty, then $U \cap V$ is empty and hence $U \cap V \in T$.

    (b) If U and V are both non-empty, then U^c and V^c are both compact. We have $(U \cap V)^c = U^c \cup V^c$. Hence $(U \cap V)^c$ is compact as it is a finite union of compact sets. Hence $U \cap V \in T$.

    Now let $(U_i)_{i \in I}$ be an arbitrary collection of elements of T. We need to show that $\bigcup_{i \in I} U_i \in T$. As before, we need to discuss two cases.

    (a) If all U_i are empty, then so is their union $\bigcup_{i \in I} U_i$ yielding $\bigcup_{i \in I} U_i \in T$.

    (b) If at least one U_j $(j \in I)$ is non-empty, then U_j^c is compact. Moreover,

$$\left(\bigcup_{i \in I} U_i \right)^c = \underbrace{\bigcap_{i \in I} U_i^c}_{\text{compact in usual } \mathbb{R} \text{ hence closed}} \subset \underbrace{U_j^c}_{\text{compact}}.$$

This implies that $\left(\bigcup_{i\in I} U_i\right)^c$ is compact, i.e. $\bigcup_{i\in I} U_i \in T$.

Thus we have proved that T is a topological space.

(2) No, T cannot be separated as no two open sets of T can be disjoint. To see this, take any two (non-empty) open sets U and V of T. If $U \cap V = \varnothing$, then $U^c \cup V^c = \mathbb{R}$ where U^c and V^c are compact in standard $\mathbb{R}$ and hence they must be bounded. Hence their (finite!) union cannot be equal to $\mathbb{R}$. Thus, T is not Hausdorff.

(3) We need to find a dense subset of $\mathbb{R}$ (density with respect to T) which is countable. The desired set is $\mathbb{Q}$. It is of course countable (and this has nothing to do with topology). It is also dense in $\mathbb{R}$. To see this, take any $x \in \mathbb{R}$. We need to show that

$$\forall U \in \mathcal{V}(x) : \ U \cap \mathbb{Q} \neq \varnothing.$$

If U is an open set (containing x), then U^c is compact with respect to standard $\mathbb{R}$ and hence it is closed with respect to standard $\mathbb{R}$. This tells us that U is open in standard $\mathbb{R}$ and so U may be written as a union of open intervals in $\mathbb{R}$ and hence $U \cap \mathbb{Q} \neq \varnothing$. This means that $\mathbb{R} \subset \overline{\mathbb{Q}}$. The other inclusion being obvious, we conclude that $\mathbb{Q}$ is dense in $\mathbb{R}$. Thus $(\mathbb{R}, T)$ is separable.

(4) Let us show that $\mathbb{R}$ is compact in this topology. Let $\mathcal{U} = \{U_i\}_{i\in I}$ be an open cover of $\mathbb{R}$ in X, i.e. $\mathbb{R} \subset \bigcup_{i\in I} U_i$ and all U_i^c are compact in $\mathbb{R}$. But we observe that for some $j \in I$

$$U_j^c \subset \mathbb{R} \subset \bigcup_{i\in I} U_i.$$

Since all U_i are open in $\mathbb{R}$ (why?), the compact U_j^c is covered by $\mathcal{U} = \{U_i\}_{i\in I}$ and hence $U_j^c \subset \bigcup_{i=1}^{p} U_i$. Thus

$$\mathbb{R} = U_j \cup U_j^c \subset U_j \cup \bigcup_{i=1}^{p} U_i \ \text{(a finite subcollection of } \mathcal{U}\text{)}.$$

This establishes the compactness of $\mathbb{R}$ in T.

SOLUTION 5.3.14. We recall that the topology T was defined on $[-1, 1]$ as

$$U \text{ open in } T \iff \{0\} \not\subset U \text{ or } (-1, 1) \subset U.$$

Let us show that $[-1, 1]$ is compact in T. Let $\mathcal{U} = \{U_i\}_{i \in I}$ be an open cover of $[-1, 1]$, i.e. $[-1, 1] \subset \bigcup_{i \in I} U_i$. Since $0 \in [-1, 1]$, this implies the existence of a j_0 in I such that $\{0\} \subset U_{j_0}$. By definition of this topology, this forces us to have $(-1, 1) \subset U_{j_0}$. Similarly,

$$\exists j_1, j_2 \in I : \{-1\} \subset U_{j_1} \text{ and } \{1\} \subset U_{j_2}.$$

Thus,

$$[-1, 1] = \{-1\} \cup (-1, 1) \cup \{1\} \subset U_{j_1} \cup U_{j_0} \cup U_{j_2},$$

i.e. proving the compactness of $[-1, 1]$ as required.

SOLUTION 5.3.15. Let $x_n = \frac{1}{n}$ $(n \geq 2)$ define a sequence in $(0, 1)$. It obviously converges to 0 and hence so do all its subsequences! But $0 \notin (0, 1)$ and hence $(0, 1)$ is not sequentially compact. Hence $(0, 1)$ is not compact.

SOLUTION 5.3.16. We know that the projections

$$p : X \times Y \to X$$
$$(x, y) \mapsto p(x, y) = x$$

and

$$q : X \times Y \to X$$
$$(x, y) \mapsto q(x, y) = y$$

are continuous functions (see Exercise 4.3.30). Since $X \times Y$ is compact and p is *continuous* and *onto*, $p(X \times Y) = X$ is compact. An analogous reasoning show that Y is also compact.

SOLUTION 5.3.17.

(1) That A is bounded is clear. Let us show that it is closed. We can write

$$A = \{x \in \mathbb{Q} : \sqrt{2} < x < \sqrt{3}\} \cup \{x \in \mathbb{Q} : -\sqrt{3} < x < -\sqrt{2}\}.$$

Call the first set in the union A_1 and the second A_2. Now A_1 is closed in $\mathbb{Q}$ since one can write

$$A_1 = [\sqrt{2}, \sqrt{3}] \cap \mathbb{Q}.$$

Also A_2 is closed in $\mathbb{Q}$ and hence A is closed in $\mathbb{Q}$. To show that A is not compact, we can consider the open covering $\mathcal{U} = \{U_n\}_{n \geq 1}$ where

$$U_n = \left\{x \in \mathbb{Q} : 2 + \frac{1}{n} < x^2 < 3 - \frac{1}{n}\right\}$$

and we can easily verify that it has no finite subcover.

(2) Finally, yes A is open in $\mathbb{Q}$ as one can write

$$A = [(\sqrt{2}, \sqrt{3}) \cup (-\sqrt{3}, -\sqrt{2})] \cap \mathbb{Q}.$$

SOLUTION 5.3.18.

(1) $(\mathbb{R}, d)$ is bounded if there exists an $r > 0$ such that $\mathbb{R} \subset B_c(0, r)$. This "$r$" is 1 since

$$\forall x \in \mathbb{R} : d(x, 0) \leq 1.$$

Thus $(\mathbb{R}, d)$ is bounded.

(2) Let us show that $(\mathbb{R}, d)$ is not sequentially compact. We show that (a_n) does not have any convergent subsequence. Before that we obviously have

$$d(a_n, a_m) = d(n, m) = \inf(|n - m|, 1) = \begin{cases} 0, & n = m, \\ 1, & n \neq m \end{cases}$$

($d(n, m)$ is worth 1 in the case $n \neq m$ since n and m are *integers*). In order to reach a contradiction, assume that $(a_{n(k)})$ converges to a real number, say a. Hence

$$\forall \varepsilon > 0, \exists K \in \mathbb{N}, \forall k \ (k \geq K \Rightarrow d(a_{n(k)}, a) < \varepsilon).$$

In particular for $\varepsilon = \frac{1}{4}$, there is $K \in \mathbb{N}$ such that

$$d(a_{n(k)}, a) < \frac{1}{4}.$$

whenever $k \geq K$. Hence

$$d(a_{n(K)}, a_{n(K+1)}) \leq d(a_{n(K)}, a) + d(a, a_{n(K+1)}) < \frac{1}{4} + \frac{1}{4} = \frac{1}{2}.$$

Now, remember that $k \mapsto n(k)$ is strictly increasing and hence $n(K+1) > n(K)$ (as $K+1 > K$) and hence $n(K+1) \neq n(K)$. Therefore,

$$d(n(K + 1), n(K)) = 1 \not< \frac{1}{2},$$

proving the non-compactness of $\mathbb{R}$.

(3) We have just seen that $\mathbb{R}$ was bounded with respect to d. Since it is the whole set, then it is closed. This is yet another example of a closed and bounded set which is not compact.

SOLUTION 5.3.19.

(1) Set $x_n = n$ for each $n \in \mathbb{N}$. Then (x_n) is a sequence in $\mathbb{R}$ from which no convergent subsequence can be extracted for

$$\lim_{n \to \infty} \delta(n, x) = \lim_{n \to \infty} |\arctan n - \arctan x| = \left| \frac{\pi}{2} - \arctan x \right| > 0$$

for all $x \in \mathbb{R}$. Thus $\mathbb{R}$ is not sequentially compact.

(2) This another example of a closed and bounded set which is not compact (that $\mathbb{R}$ is bounded was done in Exercise 2.3.26 and that $\mathbb{R}$ is closed is clear).

SOLUTION 5.3.20.

(1) We know from a basic algebra course that $\dim X = \infty$.
(2) The unit closed ball is not sequentially compact in $C([0,1],\mathbb{R})$. To see this, let (f_n) be the moving bump (see Exercise 4.3.27). Then for all different n, m

$$d_\infty(f_n, f_m) = 1.$$

Since $f_n \in B_c(0,1)$ for all n, we deduce from the equality just above that (f_n) is a sequence in the closed unit ball which cannot have a convergent subsequence. Thus the closed unit ball in $C([0,1],\mathbb{R})$ is not sequentially compact.
(3) Since we are in a metric space, the closed unit ball is not compact either.
(4) No, it is not locally compact. The proof relies on results from normed vector spaces, so we shall not include it here.

SOLUTION 5.3.21.

(1) No, $f(\mathbb{R}) = \mathbb{R}^+$ is not compact (see Exercise 5.3.2).
(2) Since $\mathbb{R}$ is compact in X and $f(\mathbb{R})$ is not compact, we deduce that f cannot be continuous.

SOLUTION 5.3.22. The answer is no in both cases, that is, $[0,1]$ is neither homeomorphic to $[0,\infty)$ nor to $(0,1]$. The reason is that $[0,1]$ is compact and the other two sets are not. Hence no homeomorphism between the two sets exists.

SOLUTION 5.3.23. Since the intersection of all the A_n is contained in each A_n we have for any f

$$f\left(\bigcap_{n\in\mathbb{N}} A_n\right) \subset \bigcap_{n\in\mathbb{N}} f(A_n).$$

To prove the reverse inclusion, we first observe that since X is compact, the hypotheses on A_n guarantee that $\bigcap_{n\in\mathbb{N}} A_n \neq \varnothing$ and hence $f\left(\bigcap_{n\in\mathbb{N}} A_n\right) \neq \varnothing$. This yields

$$\bigcap_{n\in\mathbb{N}} f(A_n) \neq \varnothing \text{ since } f\left(\bigcap_{n\in\mathbb{N}} A_n\right) \subset \bigcap_{n\in\mathbb{N}} f(A_n).$$

Let $y \in \bigcap_{n \in \mathbb{N}} f(A_n)$. Then $y \in f(A_n)$ for all n. Hence for all n, $y = f(x_n)$ with $x_n \in A_n \subset X$. Therefore, (x_n) being a sequence in X which is compact or equivalently sequentially compact, it possesses a convergent subsequence, denoted by $(x_{n(k)})$. Let $x \in X$ be its limit.

Since f is continuous or equivalently sequentially continuous, we have

$$\lim_{k \to \infty} f(x_{n(k)}) = f(x) = y \text{ (why?)}$$

The proof will be complete as soon as we verify that x is in each A_n for all n. To see this, fix an $m \in \mathbb{N}$. Then

$$x_n \in A_n \subset A_m, \ \forall n \geq m$$

for (A_n) is decreasing. A fortiori, the previous holds true for each k. Passing to the limit (as $k \to \infty$) we immediately observe that x must be in $\overline{A_m} = A_m$ (why?) and this is for each m. Thus

$$x \in \bigcap_{m \in \mathbb{N}} A_m = \bigcap_{n \in \mathbb{N}} A_n.$$

Thus the proof is complete.

SOLUTION 5.3.24.

(1) (a) It is clear that

$$\forall x \in X, \ \forall U \in \mathcal{V}(x) : \ U \subset X.$$

Since X is compact, we see immediately that X is locally compact.

REMARK. The converse is obviously not always true. See the next answer.

(b) Let $x \in \mathbb{R}$. In usual $\mathbb{R}$, $[x - 1, x - 1]$ is a compact set. It contains $(x - 1, x - 1)$ which is a neighborhood of x. Therefore, $\mathbb{R}$ is locally compact (remember that it is not compact!).

(c) If $\mathbb{Q}$ were locally compact, there would exist a compact set V such that a neighborhood of 0 (for example!) U, say, would satisfy $U \subset V$. But $U = (-r, r) \cap \mathbb{Q}$ for some $r > 0$. For irrational r, $[-r, r] \cap \mathbb{Q}$ would be a closed set in V and hence it would be compact. This is evidently not true for we can take a sequence of *rationals*, (r_n) say, which converges to the (irrational!) r. Thus (r_n) cannot have a converging subsequence as all subsequences would converge to r too (and $r \notin [-r, r] \cap \mathbb{Q}$!).

If r is not irrational, then replace $[-r, r]$ by $[-s, s]$ where $s < r$ is irrational and follow a similar reasoning!

(d) A similar proof as before applies to show that $\mathbb{R} \setminus \mathbb{Q}$ is not locally compact.

(e) Let $x \in X$. Then $\{x\}$ is a compact set (why?) and it is a neighborhood of x. The proof is complete.

(f) Let $x \in X$. It is clear that $\{a, x\}$ is compact and a neighborhood of x simultaneously. Thus X is locally compact.

(2) From the previous answers, $\mathbb{Q}$ is Hausdorff but it is not locally compact. Also, $\mathbb{R}$ is Hausdorff and locally compact.

SOLUTION 5.3.25. Let f be the identity mapping from X onto Y where X is $\mathbb{R} \setminus \mathbb{Q}$ endowed with the discrete topology and Y is $\mathbb{R} \setminus \mathbb{Q}$ equipped with the usual topology.

Then f is continuous and we have $f(X) = Y$. Now from Exercise 5.3.24, X is locally compact while Y is not.

SOLUTION 5.3.26.

(1) Let C be a closed set in X. Let $x \in C$. Let Y be a compact set that contains a neighborhood U of x. Then $Y \cap C$ is closed in C and thus it is compact. Besides

$$x \in U \cap C \subset Y \cap C$$

and so $U \cap C$ is a neighborhood of x in C, proving the local compactness of C.

(2) Let U be an open set in X and let $x \in U$. From the hint, there exists $V \in \mathcal{V}(x)$ such that $\overline{V}$ is compact and $\overline{V} \subset U$. Since $V \subset \overline{V}$, U becomes locally compact.

(3) Well, $\{(0, 0)\}$ is compact and hence locally compact. The set $\{(x, y) \in \mathbb{R}^2 : x > 0\}$ is locally compact by the previous question for it is open in usual $\mathbb{R}^2$ (which is Hausdorff!).

(4) The given set is not locally compact. It is not locally compact at 0 (which is the only point causing the problem of non-local compactness) as the reader may check.

(5) We deduce from the previous two questions that the union of two locally compact spaces need not remain locally compact.

SOLUTION 5.3.27.

(1) No! For example, take $X = [0, 1]$ and $f(x) = \frac{x^2}{3}$.

(2) Define a function

$$\varphi : (X, d) \to (\mathbb{R}, |\cdot|) \text{ by } \varphi(x) = d(x, f(x)).$$

We claim that φ is continuous on X. To see this, let $x, y \in X$. Then we have by Exercise 2.5.2

$$|\varphi(y) - \varphi(x)| \le d(x, y) + d(f(x), f(y)) \le 2d(x, y)$$

(for all $x, y \in X$). Therefore, φ is continuous on X which is compact and so the function φ attains its minimum (and its maximum too!) at some $a \in X$. Let us show that a is the looked for fixed point. Assume $f(a) \ne a$. Then

$$\varphi(f(a)) = d(f(a), (f \circ f)(a)) < d(a, f(a)) = \varphi(a)$$

and we realize that φ does not attain its minimum at a anymore! Therefore, $f(a) = a$, showing the f admits a fixed point. Let us now show its uniqueness.

Assume there were two *different* fixed points x and y say, i.e. $f(x) = x$ and $f(y) = y$ with $x \ne y$. Then

$$d(x, y) = d(f(x), f(y)) < d(x, y)$$

which is impossible and hence $x = y$ necessarily. The proof is complete.

REMARK. An instance showing the importance of the compactness hypothesis may be found in Exercise 7.3.20.

SOLUTION 5.3.28. Let (X, d) be a compact metric space. Then it is not hard to see that $\mathcal{U} = \bigcup_{x \in X} B(x, \frac{1}{n})$ constitutes an open cover for X. By compactness, we can cover X by $\bigcup_{k=1}^{m} B(x_k, \frac{1}{n})$ which we denote by S_n. Now set $S = \bigcup_{n \in \mathbb{N}} S_n$. Then it is plain that S is countable. It remains to verify that S is dense in X. Observe that

$$\forall x \in X, \ \forall n \in \mathbb{N}, \ \exists y \in S_n \subset S : \ d(x, y) < \frac{1}{n}$$

which implies that

$$d(x, S) = \inf_{s \in S} d(x, s) < \frac{1}{n}$$

for all $n \in \mathbb{N}$. Thus $d(x, S) = 0$ and hence Exercise 3.3.33 implies that $x \in \overline{S}$ which completes the proof.

SOLUTION 5.3.29. The answer depends on the cardinality of X.

(1) If $\text{card} X$ is finite, then (X, d) is totally bounded (see the "True/False" section).

(2) If card$X$ is infinite, then let $0 < \varepsilon < \frac{1}{3}$ (for instance). Then the only subsets of X whose diameter is smaller than ε are the singletons $\{x\}$, where $x \in X$ (and evidently $\varnothing$). It thus becomes clear that it is impossible to cover the *infinite* X by a *finite* number of singletons! This tells us that (X, d) is not totally bounded and that only finite discrete metric spaces are totally bounded.

5.4. Hints/Answers to Tests

SOLUTION 35. Well, do it!...

SOLUTION 36. Yes $\{a\}$ is compact since it is finite. X is not compact (if it is infinite!). To see this we observe that $\{a, x\}$ are open sets for each $x \in X$ (why?) whose union covers X from which no finite subcollection can cover the whole of X.

We deduce from that $\{a\}$ is not relatively compact since from Exercise 3.3.25, $\overline{\{a\}} = X$.

SOLUTION 37. It suffices to show $S \subset T$. If V is a closed set in S, then V is compact and compact sets in Hausdorff spaces are closed...Alternatively, use the identity mapping between S and T...

SOLUTION 38. Consider the open cover consisted of the sets $\{1\}, \{1, 2\} \cdots, \{1, 2, \cdots, n\}, \cdots$ which is not reducible to a finite cover...

SOLUTION 39. No, $\mathbb{R}$ is not compact in $\mathbb{R}_\ell$. One reason is that $\{[-n, n)\}_{n \in \mathbb{N}}$ is an open cover of $\mathbb{R}$ and the rest is obvious...

SOLUTION 40. Yes...

SOLUTION 41. No! (Why?)...

SOLUTION 42. Fairly simple...

SOLUTION 43. Yes (why?)...

SOLUTION 44. Yes by Heine's theorem!...

SOLUTION 45. Yes! Why?...

CHAPTER 6

Connected Spaces

6.2. True or False: Answers

ANSWERS.

(1) If (X, T) is not connected, then (X, T') is not connected. The reason is simple, that is, if U and V are open in T such that $U \cup V = X$ and $U \cap V = \varnothing$ (meaning that (X, T) is not connected), then U and V are open in T' with $U \cup V = X$ and $U \cap V = \varnothing$ as well, leading to the non-connectedness of (X, T'). The other implication does not always hold. As a counterexample, take usual $\mathbb{R}$ and the discrete $\mathbb{R}$, denoted by $\mathbb{R}_{dis}$. Then $\mathbb{R}$ is connected while $\mathbb{R}_{dis}$ is not and yet $\mathbb{R} \subset \mathbb{R}_{dis}$.

(2) The answer is no! In usual $\mathbb{R}$, $\{1, 2\}$ is closed and not connected (it is not an interval) but $\mathbb{R}$ is connected.

(3) The answer is no! For instance, in the usual topology, $A = [-1, 1]$ is connected while its boundary $\{-1, 1\}$ is not.

(4) The answer is no! A counterexample is $\mathbb{Q}$ which is not connected (as will be seen below) while $\overline{\mathbb{Q}} = \mathbb{R}$ is connected. The converse is always true, i.e. if A is connected, so is its closure $\overline{A}$. See Exercise 6.3.7.

(5) The answer is no! For example, in the usual topology, both $A = [0, 1]$ and $B = [2, 3]$ are connected while their union is not.

(6) The answer is no! Let A be the unit circle (in $\mathbb{R}^2$ endowed with its euclidian metric) and let B be the real axis. They are both connected whilst their intersection, equal to $\{-1, 1\}$, is not connected.

(7) The answer is no in general. Take $B = B_c((0, 1), 1) \cup B_c((0, -1), 1)$. Then B is connected (see Exercise 6.3.1). It can be shown that $\overset{\circ}{B} = B((0, 1), 1) \cup B((0, -1), 1)$ which is not connected (see again Exercise 6.3.1).

It is worth stating that the result is true in the usual topology of $\mathbb{R}$. It is known that the connected parts of standard $\mathbb{R}$

are the intervals and the interior of an interval in $\mathbb{R}$ remains an interval and hence it is connected.

(8) If both $[-1,1]$ and $\mathbb{R}$ are endowed with the usual topology, the answer is yes. But if we just endow $[-1,1]$ with the discrete topology (and leave $\mathbb{R}$ equipped with the usual topology), then the given statement becomes false. Consider the function f defined by

$$f(x) = \begin{cases} -1, & -1 \le x \le 0, \\ 1, & 0 < x \le 1. \end{cases}$$

The discreteness of the topology of $[-1,1]$ guarantees the continuity of f. Besides, f verifies $f(-1)f(1) < 0$ and observe that $\forall x \in [-1,1]: f(x) \ne 0$.

(9) True (in usual $\mathbb{R}$)! Polynomials on $\mathbb{R}$ are continuous. Since the degree is odd, then the limits at $\pm\infty$ are $\pm\infty$ or $\mp\infty$. Thus the polynomial in question will have to pass through the real axis at least once.

(10) No! In usual $\mathbb{R}$, it will be shown in Exercise 6.3.11 that $\mathbb{R}^*$ is not path-connected. However, $\overline{\mathbb{R}^*} = \mathbb{R}$ is path-connected.

(11) The answer is no! See Exercise 6.3.15.

(12) The answer is no! Consider the identity map from X into Y (denoted by f) where X is the usual topology of $\mathbb{R}$ and Y is $\mathbb{R}$ equipped with the co-finite topology. Then f is continuous. Then, $[0,1] \cup [2,3]$ is connected in Y (cf. Exercise 6.3.1 below) whereas

$$f^{-1}([0,1] \cup [2,3]) = [0,1] \cup [2,3]$$

is not a connected set in X as it is not an interval. The answer is also no in usual $\mathbb{R}$. Take $f : \mathbb{R} \to \mathbb{R}$ defined by $f(x) = x^2$. Then $\{2\}$ is connected but $f^{-1}(\{2\}) = \{\sqrt{2}, -\sqrt{2}\}$ is not connected.

(13) True! Let $f : X \to Y$ be such a homeomorphism. Let $x \in X$ and let C_x be its component. We ought to show that

$$C_{f(x)} = f(C_x).$$

(a) Since C_x is connected, $f(C_x)$ is connected for f is continuous. But $f(x) \in f(C_x)$. Thus $f(C_x) \subset C_{f(x)}$.

(b) Conversely, assume that $C_{f(x)} \subset A$ for some connected $A \ne f(C_x)$. Then $f^{-1}(C_{f(x)}) \subset f^{-1}(A)$. By the proof of the other part,

$$C_x = f^{-1}[f(C_x)] \subset f^{-1}(C_{f(x)}) \subset f^{-1}(A).$$

But f^{-1} is continuous and hence $f^{-1}(A)$ is connected. Since $x \in f^{-1}(A)$, we see that we arrived at a contradiction for C_x is the largest connected set that contains x! The proof is over.

REMARK. Thanks to this result, we can prove interesting results on the non-homeomorphy of some spaces. See some exercises below.

(14) True! In fact, this a simple corollary of the fact that the continuous image of a connected set (path-connected respectively) is connected (path-connected respectively).
(15) True! Both implications are trivial. The right-to-left implication is the most trivial. For the other one, if X is connected, then it is obviously the largest connected set in X containing x!
(16) No, there is no contradiction between this question and the closedness of the components. First $\mathbb{R}^*$ is not connected since it has two components (this is one proof among others). Of course both $(0, \infty)$ and $(-\infty, 0)$ are not closed in $\mathbb{R}$ but they are closed in $\mathbb{R}^*$ with the induced topology (which is the meant topology in the definition).

6.3. Solutions to Exercises

SOLUTION 6.3.1.

(1) The only open and closed parts simultaneously in T are A and $\varnothing$. Hence A is connected.
(2) The discrete topology is disconnected since every subset is clopen.
(3) Let $T = \{\varnothing\} \cup \{U \subset \mathbb{R} : U^c \text{ finite }\}$. Then $\mathbb{R}$ is connected in T. If $\mathbb{R}$ were not connected in this topology, then there would exist two non-empty and open sets U and V in T (i.e. U^c and V^c are both finite) such that

$$\begin{cases} U \cup V = \mathbb{R}, \\ U \cap V = \varnothing. \end{cases}$$

Since $U \cap V = \varnothing$, $U^c \cup V^c = \mathbb{R}$ and this clearly contradicts the infiniteness of $\mathbb{R}$!
(4) Let $A = B((0,1),1) \cup B((0,-1),1)$. The intersection of the two balls is empty as we will show. Let $(x,y) \in B((0,1),1) \cup$

$B((0, -1), 1)$. Then

$$\begin{cases} x^2 + (y-1)^2 < 1, \\ x^2 + (y+1)^2 < 1 \end{cases} \text{which gives} \begin{cases} x^2 + y^2 - 2y < 0, \\ x^2 + y^2 + 2y < 0 \end{cases}$$

and hence $x^2 + y^2 < 0$ which clearly cannot occur for any real couple (x, y). Thus A is non connected since $\{B((0,1),1), B((0,-1),1)\}$ is an open partition.

(5) The given set A is connected since it is the union of two connected sets whose intersection is not empty (it is equal to $\{(0,0)\}$).

(6) A is also connected. From the penultimate proposition of Subsection 6.1.1, we know that if X and Y are two non-empty connected sets such that $\overline{X} \cap Y \neq \varnothing$, then $X \cup Y$ is connected. Now, we have (as we are in Euclidian $\mathbb{R}^2$)

$$\overline{B((0,1),1)} \cap B_c((0,-1),1) = B_c((0,1),1) \cap B_c((0,-1),1) = \{(0,0)\} \neq \varnothing.$$

Thus $A = B((0,1),1) \cup B_c((0,-1),1)$ is in effect connected.

SOLUTION 6.3.2. First, we can write

$$A = \{M \in \mathcal{M}_n(\mathbb{R}) : \det M \neq 0\}$$

which can also be written as

$$A = \underbrace{\{M \in \mathcal{M}_n(\mathbb{R}) : \det M < 0\}}_{A_1} \cup \underbrace{\{M \in \mathcal{M}_n(\mathbb{R}) : \det M > 0\}}_{A_2}.$$

But, the function "det" is real-valued and continuous (see Exercise 4.3.9). Hence, A_1 and A_2 are open (why?). Lastly, it is plain that $A_1 \cap A_2 = \varnothing$. Thus, $\{A_1, A_2\}$ is an "open partition" of A which means that A cannot be connected.

SOLUTION 6.3.3. We give three proofs

- First proof: $\mathbb{Q}$ is not connected since it is not an interval. If it were one, then for any $x, y \in \mathbb{Q}$, one would have $(x, y) \subset \mathbb{Q}$. But this is obviously not true. For instance, $2, 4 \in \mathbb{Q}$ but $(2, 4) \not\subset \mathbb{Q}$ as $\pi \in (2, 4)$ while $\pi \notin \mathbb{Q}$.
- Second proof: $\mathbb{Q}$ is not connected since there is another open and closed set in the subspace topology of $\mathbb{Q}$ (apart from $\mathbb{Q}$ itself and $\varnothing$). Take $A = \mathbb{Q} \cap (-\infty, \sqrt{2})$. Then A is clopen in the subspace topology of $\mathbb{Q}$ for

$$A = \mathbb{Q} \cap (-\infty, \sqrt{2}) = \mathbb{Q} \cap (-\infty, \sqrt{2}]$$

and obviously $(-\infty, \sqrt{2})$ is open in $\mathbb{R}$ and $(-\infty, \sqrt{2}]$ is closed in $\mathbb{R}$.

- Third proof: The following two sets $A = \mathbb{Q} \cap (-\infty, \sqrt{2})$ and $B = \mathbb{Q} \cap (\sqrt{2}, \infty)$ are open in $\mathbb{Q}$ in the subspace topology and constitute an open partition for $\mathbb{Q}$.

SOLUTION 6.3.4.

(1) As for $\mathbb{R} \backslash \mathbb{Q}$ one can proceed exactly as in the foregoing exercise to show that this set is not connected.
(2) The set $\{\frac{1}{n} : n \geq 1\}$ is not connected since it is not an interval, say.
(3) No, since $[0, 1) \cup (1, 2)$ is not an interval.
(4) $\mathbb{N}$ is not connected since

$$U = \left(0, \frac{3}{2}\right) \cap \mathbb{N}, \; V = \left(\frac{3}{2}, +\infty\right) \cap \mathbb{N}$$

are both open in $\mathbb{N}$ and they constitute a partition for $\mathbb{N}$.

SOLUTION 6.3.5. Assume $[0, 2)$ is not connected, then it could be written as $[0, 2) = U \cup V$ with $U \cap V = \varnothing$ where U and V are open. But $U \cap V = \varnothing$ does not hold since every non-empty open set in this topology contains at least 0. Thus, $[0, 2)$ is connected with respect to the given topology.

SOLUTION 6.3.6.

(1) We have $z \in A$ iff $\operatorname{Im} z \neq 0$. Hence there are two components corresponding to the cases $\operatorname{Im} z > 0$ and $\operatorname{Im} z < 0$.
(2) The components of A are

$$B = \{(x, y) \in \mathbb{R}^2 : x < y\} \text{ and } C = \{(x, y) \in \mathbb{R}^2 : x > y\}.$$

(3) A has two components, to wit $B((0, 1), 1)$ and $B((0, -1), 1)$. B is connected and hence it has one component, viz itself.

SOLUTION 6.3.7.

(1) First, we recall that $\overline{A}^B = B \cap \overline{A}^X$. Since $B \subset \overline{A}^X$, we get that $\overline{A}^B = B$. Now, let $f : B \to \{0, 1\}$ be a continuous function. We ought to show that f is constant. Since $A \subset B$, we deduce that f is continuous on A too. By the connectedness of A, we know that f must be constant, i.e. $f(A) = \{0\}$ or $f(A) = \{1\}$. Assume $f(A) = \{1\}$. The continuity of f and the closedness of $\{1\}$ lead to

$$f(B) = f(\overline{A}^B) \subset \overline{f(A)} = \overline{\{1\}} = \{1\}.$$

Thus $f(B) = \{1\}$, i.e. f is constant and hence B is connected. The case $f(A) = \{0\}$ is very analogous to the foregoing one and hence we leave it to the reader.

(2) Since $A \subset \overline{A} \subset \overline{A}$, the previous result applies to $\overline{A}$, i.e. $\overline{A}$ is connected whenever A is connected.

SOLUTION 6.3.8.

(1) It is clear that $(-\infty, 0)$ inherits the usual topology as a subspace of $\mathbb{R}_K$ and so does $(0, \infty)$. Thus, both spaces are connected.

(2) Since $(-\infty, 0)$ and $(0, \infty)$ are connected, so are their closures, i.e. $(-\infty, 0]$ and $[0, \infty)$. But

$$(-\infty, 0] \cap [0, \infty) = \{0\} \neq \varnothing$$

and so $(-\infty, 0] \cup [0, \infty) = \mathbb{R}_K$ is connected.

SOLUTION 6.3.9. Assume that such function exists. Since $\mathbb{Q}$ is countable, so is $f(\mathbb{Q})$ by Exercise 1.3.7. The set $f(\mathbb{R} \setminus \mathbb{Q})$ is also countable, but this time, since it is a subset of $\mathbb{Q}$. We hence end up with $f(\mathbb{R})$ being countable.

Now, $\mathbb{R}$ is connected and so is its direct image $f(\mathbb{R})$ by the continuity of f. Thus $f(\mathbb{R})$ must be an interval. But, it is clear that the only countable intervals are the very particular intervals $[a, a]$, which are singletons $\{a\}$ ($a \in \mathbb{R}$). This forces f to be constant and hence $f(x) = a$ for all $x \in \mathbb{R}$.

Going back to the hypotheses again we see that

$$f(\mathbb{Q}) \subset \mathbb{R} \setminus \mathbb{Q} \Rightarrow a \in \mathbb{R} \setminus \mathbb{Q}$$

and

$$f(\mathbb{R} \setminus \mathbb{Q}) \subset \mathbb{Q} \Rightarrow a \in \mathbb{Q},$$

which is a clear contradiction. Thus no such function f verifying the hypotheses exists.

SOLUTION 6.3.10. If $\mathbb{N}$ were path-connected, then there would exist a continuous mapping $f : [0, 1] \to \mathbb{N}$. Now we have

$$\mathbb{N} = \bigcup_{n \in \mathbb{N}} \{n\} \implies [0, 1] = f^{-1}(\mathbb{N}) = f^{-1}(\bigcup_{n \in \mathbb{N}} \{n\}) = \bigcup_{n \in \mathbb{N}} f^{-1}(\{n\}).$$

Since f is continuous and the singletons $\{n\}$ are closed for each n (they are finite), the sets $f^{-1}(\{n\})$ are closed. Hence, we would end up with a countable union of disjoint and closed sets equal to $[0, 1]$ which is continuum, i.e. compact, Hausdorff and connected. But this clearly contradicts the Sierpinski theorem which says that no continuum can be written as a countable union of many closed sets. Thus $\mathbb{N}$ is not path-connected with respect to the co-finite topology.

REMARK. The Sierpinski theorem used above was proved in 1918. Its original proof may be found in [**12**].

SOLUTION 6.3.11. Assume that $\mathbb{R}^*$ is path-connected. Since $-1, 1 \in \mathbb{R}^*$, there exists a continuous function f defined on $[0, 1]$, taking values in $\mathbb{R}^*$ such that $f(0) = -1$ and $f(1) = 1$. By the intermediate value theorem there exists at least an α in $[0, 1]$ such that $f(\alpha) = 0$ which clearly leads to a contradiction as $f(x) \neq 0$, for all $x \in [0, 1]$.

SOLUTION 6.3.12.

(1) Let us show that if C is convex, then it is path-connected. Let $x, y \in C$ and consider the function f defined on $[0, 1]$ and taking its values in C such that $f(t) = (1 - t)x + ty$.

Then f is continuous and $f(0) = x$ and $f(1) = y$. Thus C is path-connected and hence connected.

(2) The set $\mathbb{R}^n$ is obviously convex (why?) and hence it is path-connected and so it is connected too.

Now, we show that the closed ball in $\mathbb{R}^n$ is convex (the proof for the open ball is very similar and hence left to the reader). WLOG, we may do the proof for $n = 2$ only. Let $B_c((a, b), r)$ be the closed ball of center $(a, b) \in \mathbb{R}^2$ and radius $r > 0$. Let $(x, y), (x', y') \in B_c((a, b), r)$ and let $r > 0$. We need to show that $(1 - t)(x, y) + t(x', y') \in B_c((a, b), r)$, i.e.

$$[(1 - t)x + tx' - a]^2 + [(1 - t)y + ty' - b]^2 \leq r^2.$$

Details are left to the reader.

SOLUTION 6.3.13.

(1) If $n = 1$, then it is showed above that $\mathbb{R}^*$ is not path connected. If $n \geq 2$, then any two elements of $\mathbb{R}^n$ can always be joined by a segment if this segment does not pass through the origin. If it does pass through it, then we may consider another point z and draw a segment from x to z, then one from z to y.

(2) If $n = 1$, then S^{n-1} is reduced to S^0, which in its turn, is reduced to the set $\{-1, 1\}$. Hence it is not path-connected.

If $n \geq 2$, then S^{n-1} becomes connected via the following arguments. Consider the function $f : \mathbb{R}^n \setminus \{0_{\mathbb{R}^n}\} \to S^{n-1}$ defined by

$$f(x_1, x_2, \cdots, x_n) = \frac{(x_1, x_2, \cdots, x_n)}{\sqrt{(x_1^2 + x_2^2 + \cdots + x_n^2)}}.$$

Then f is continuous and onto. But $\mathbb{R}^n \setminus \{0_{\mathbb{R}^n}\}$ is path-connected (for $n \geq 2$) and hence $f(\mathbb{R}^n \setminus \{0_{\mathbb{R}^n}\}) = S^{n-1}$ is path-connected too.

Solution 6.3.14.

(1) The functions f and g are obviously continuous on their domains since their components are so. Now $f \circ g : A \to A$ and $g \circ f : [a, b] \times \mathbb{T} \to [a, b] \times \mathbb{T}$ are given respectively by

$$(f \circ g)(x, y) = (x, y) \text{ and } (g \circ f)(t, (x, y)) = (t, (x, y)).$$

Hence f is a continuous bijection. Its inverse g is also continuous. Thus f is a homeomorphism.

(2) Since $[a, b]$ and $\mathbb{T}$ are both path-connected, so is $[a, b] \times \mathbb{T}$. Thus A is, in its turn, path-connected.

Solution 6.3.15.

(1) First, $\overline{f}$ is continuous since it is the restriction of f. So is $(\overline{f})^{-1}$ too as it is the restriction of f^{-1}. Now, it is clear that $\overline{f}$ is also a bijection. Therefore, $\overline{f}$ is a homeomorphism.

(2) Assume that $f : \mathbb{R} \to \mathbb{R}^2$ is a homeomorphism. According to the previous question, $\overline{f} : \mathbb{R} \setminus \{0\} \to \mathbb{R}^2 \setminus \{f(0)\}$ should remain a homeomorphism. But $(\overline{f})^{-1}$ is continuous, $\mathbb{R}^2 \setminus \{f(0)\}$ is path-connected (cf. Exercise 6.3.13) and $\mathbb{R}^*$ is not path-connected (see Exercise 6.3.11). This a clearly a contradiction since the image of a path-connected set under a continuous function has to stay path-connected. Thus $\mathbb{R}$ and $\mathbb{R}^2$ are not homeomorphic.

Solution 6.3.16. X and S are not homeomorphic. If there were a homeomorphism between X and S, then $\overline{f} : X \setminus \{a\} \to S \setminus \{f(a)\}$ ($a \in$ X) would remain one. But if a is the junction point of X, then $X - \{a\}$ has *four* components while S deprived of one point has either one or two components depending on its position on S.

A similar reasoning shows that E and W cannot be homeomorphic.

6.4. Hints/Answers to Tests

Solution 46. Well do it!...For a two-point set find a counterexample...

Solution 47. No! One reason is that $[0, 1)$ is clopen in $\mathbb{R}_\ell$ (cf. Exercise 3.3.29)...

Solution 48. Any subset A ($a, b \in A$) or $\mathbb{R}_\ell$ is not connected as $A \cap [a, b)$ is a *proper* clopen set of A...

Solution 49. Well, $\mathbb{R}$ is connected in the co-countable topology...

As for $\mathbb{Q}$, it is not connected (Is $\mathbb{Q}$ connected in the discrete topology?)...

SOLUTION 50. The result follows from Exercise 6.3.7...

SOLUTION 51. We already know that $\mathbb{R}$ is connected in this topology...

SOLUTION 52. No! Deprive $[-1, 1]$ of one point (not 1 and not -1)...

CHAPTER 7

Complete Metric Spaces

7.2. True or False: Answers

ANSWERS.

(1) We **cannot** take $\varepsilon \geq 0$! But we can take $d(x_n, x_m) \leq \varepsilon$ with no problem at all.

(2) Let X be a a metric space. To establish its completeness, we take an *arbitrary* Cauchy sequence (x_n) in X and we show that this sequence converges to a limit x which *belongs* to X.

> **REMARK. Do not** take a particular Cauchy sequence and show it converges! The sequence must be arbitrary.

(3) On the contrary to the previous answer, if we want to show that a given metric space X is not complete, it suffices to find *one* Cauchy sequence (x_n) which *converges*. But, its limit *does not* belong to X.

Alternatively, we can show that X is not closed.

(4) If we want to use the completeness of a space X, we try to define a sequence in X and then show it is Cauchy. Now, since the space is complete, this sequence converges to a limit in X (do not say in the beginning, let (x_n) be a a arbitrary Cauchy sequence. Then it converges. Then what?)

(5) The importance is about convergence. For example, since usual $\mathbb{R}$ is complete, to prove a sequence is convergent it suffices to show it is Cauchy. For instance, to prove the series

$$\sum_{n \geq 1} \frac{1}{n^a}, \ a > 1, \text{ is convergent it suffices to show it is Cauchy in-}$$

stead of trying to find its limit (whose exact value is unknown anyway, except for $a = 2$)!

Another major point, still about convergence, is the fixed point theorem. Many solutions to problems can be obtained via the fixed point theorem (approximation problems, ordinary differential equations, integral equations...etc).

(6) False! As a "lazy" counterexample, we can just endow $(0, \infty)$ with the discrete metric (see Exercise 7.3.2). But, there are

more interesting counterexamples. For instance, if we equip $(0, \infty)$ with the following *metric*

$$d(x, y) = |\ln x - \ln y|, \ \forall x, y > 0,$$

then $(0, \infty)$ will be complete (see Exercise 7.3.16).

Evidently, $(0, \infty)$ is not complete in the usual metric.

(7) The answer is no! For instance, in usual $\mathbb{R}$, consider $\mathbb{Q}$ and $\mathbb{R} \setminus \mathbb{Q}$. They are both dense in $\mathbb{R}$ and yet

$$\mathbb{Q} \cap (\mathbb{R} \setminus \mathbb{Q}) = \varnothing.$$

Another example (similar though!) consists of taking the set of algebraic numbers and the set of transcendental ones (both as subsets of $\mathbb{R}$).

(8) The answer is yes. For example, in $\mathbb{R}$ equipped with its usual metric, both $\mathbb{Q}$ and $\mathbb{R} \setminus \mathbb{Q}$ are not complete. But, their union, i.e. $\mathbb{R}$, is complete.

(9) Only the left-to-right implication holds. Let us show that. Since (x_n) is Cauchy, for all $\varepsilon > 0$ and in particular for $\varepsilon = 1$,

$$\exists N \in \mathbb{N}, \ \forall n, m \ (n, m \geq N \Rightarrow d(x_n, x_m) \leq 1).$$

If $a = x_N$, then for all $m \geq N$: $d(a, x_m) \leq 1$. This means that $x_m \in B_c(a, 1)$. Thus it becomes clear that

$$\{x_n : \ n \in \mathbb{N}\} \subset \underbrace{\{x_1, x_2, \cdots, x_{N-1}\}}_{\text{finite hence bounded}} \cup B_c(a, 1)$$

proving the boundedness of the sequence as its range will be included in the (finite!) union of two bounded sets.

The other implication need not be true in general. From basic real analysis, it is known that $((-1)^n)_n$ is bounded but not convergent hence not Cauchy (usual $\mathbb{R}$ is complete!).

(10) The answer is yes. Since the two metrics are equivalent, we have

$$\exists \alpha, \beta > 0, \ \forall n, m \in \mathbb{N} : \ \alpha d(x_n, x_m) \leq d'(x_n, x_m) \leq \beta d(x_n, x_m)$$

giving us the desired result.

(11) The answer is yes if one of the inequalities involved in the definition of equivalent metrics is satisfied. For example, assume that d and d' are two metrics on a set X. If *only* (for some $\alpha > 0$) the inequality

$$d(x, y) \leq \alpha d'(x, y), \ \forall x, y \in X$$

holds, then d and d' are not equivalent metrics. Nevertheless, the "Cauchyness" of a sequence (x_n) in (X, d') implies that of (x_n) in (X, d).

The answer is, however, *false* if none of the inequalities intervening in the definition of equivalent metrics is satisfied.

(12) It is false. While the fact that (f_n) is a Cauchy sequence of continuous functions can be seen in Exercise 7.3.12, the remaining part is false for a simple reason. That is, the convergence used to get f is the pointwise one, i.e. we are not using the right metric which should be d.

(13) False again. In this case the metric is the right one but the argument used is not quite right. What tells us that there is not another *continuous* function which is the d-limit of f_n? One possible answer is that we are in a metric space (hence Hausdorff) and the limit of a convergent sequence is always unique in a metric space. Digging more into this, the function f is Riemann integrable and d is not a metric on the set of Riemann integrable functions (see Exercise 2.3.12) and more particularly $d(f, g) = 0 \Rightarrow f = g$ (which is essential for proving uniqueness of limits!) is the axiom which *does not* hold.

(14) The answer is no! Let $f : \mathbb{R}_+^* \to \mathbb{R}_+^*$ be defined by $f(x) = \frac{1}{x}$ for all $x > 0$. Then f is evidently a homeomorphism. Now, let $x_n = \frac{1}{n}$. Then (x_n) is Cauchy. Nevertheless, $f\left(\frac{1}{n}\right) = n$ is not Cauchy (for example, since it is not bounded).

(15) True! The reader is asked to prove this in Exercise 7.5.9.

(16) False! There are many counterexamples. Consider the function

$$f : [1, +\infty) \to (0, 1] \text{ defined for all } x \geq 1 \text{ by } f(x) = \frac{1}{x}.$$

Then f is uniformly continuous and obviously $[1, +\infty)$ is complete (why?). But its image which is $(0, 1]$ is not complete as it is not closed.

(17) False! For a counterexample, see Exercise 7.3.16.

If, for example, two metric spaces $((X, d)$ and (X', d'), say) are isometric, i.e. there exists a one-to-one correspondence $f : (X, d) \to (X', d')$ such that

$$\forall x, y \in X : \ d'(f(x), f(y)) = d(x, y),$$

then X is complete iff X' is so. For a proof see Exercise 7.3.21.

REMARK. A slightly weaker result holds, that is, if there exists some one-to-one correspondence f between (X, d) and

(X', d') such that

$$\exists \alpha, \beta > 0, \ \forall x, y \in X : \ \alpha d(x, y) \le d'(f(x), f(y)) \le \beta d(x, y),$$

then X is complete iff X' is so.

(18) False! Two conditions are to be imposed on f and on A both in order that such an extension is guaranteed to exist. We recall the following theorem from the "What you need to know" section:

THEOREM. *Let (X, d) and (Y, d') be two metric spaces. Assume that $A \subset X$ is dense and that (Y, d') is complete. Let $f : (A, d) \to (Y, d')$ be uniformly continuous on A. Then there exists a unique function $g : (X, d) \to (Y, d')$ which is also uniformly continuous and such that $g(x) = f(x)$ for each $x \in A$.*

REMARK. The proof for uniqueness is a simple application of Exercise 4.3.23.

REMARK. In normed vector spaces, the continuity is equivalent to uniform continuity whenever the function or map is linear. Thus a very similar result holds in normed vector spaces by assuming that the function is linear and continuous.

Let us give two counterexample which show that the conditions Y complete and f uniformly continuous cannot be dispensed with.

(a) Let f be the identity function from $\mathbb{Q}$ onto $\mathbb{Q}$ in the usual topology. Then f is uniformly continuous (note that the codomain, i.e."arrival" set, $\mathbb{Q}$ is not complete). Assume that f had an extension $\tilde{f}$, then $\tilde{f}_{|\mathbb{Q}}(x) = x$. Since $\mathbb{Q}$ is dense in $\mathbb{R}$, an irrational y is a limit of a sequence $(y_n) \subset \mathbb{Q}$. We would then have

$$\tilde{f}(y) = \lim_{n \to \infty} \tilde{f}(y_n) = \lim_{n \to \infty} f(y_n) = \lim_{n \to \infty} y_n = y \notin \mathbb{Q}.$$

This illustrates the importance of Y being complete.

(b) Let $X = [0, 1]$, $A = (0, 1]$ and $Y = \mathbb{R}$. Define a function f on A as $f(x) = \ln x$. Then A is dense in X, but f is "only" continuous (the continuity is not uniform). If $\tilde{f}$ were such an extension of f, then we would have

$$\tilde{f}(0) = \lim_{x \to 0} \tilde{f}(x) = \lim_{x \to 0} f(x) = \lim_{x \to 0} \ln x = -\infty.$$

This tells us that such an extension does not exist showing the importance of the uniform continuity.

(19) There are many properties shared by compact and complete metric spaces. An important result is that *every compact metric space is complete* (the reader is asked to give a proof in Exercise 7.3.6). Some of the properties illustrating the analogy are listed below:

(a) Any closed subspace of a compact space is compact and any closed set in a complete metric space is complete.

(b) Any compact subspace of a metric space is closed and so is any complete subspace.

(c) The *finite* union of complete metric spaces is complete and a *finite* union of compact spaces is compact too.

(d) The arbitrary intersection of complete metric spaces is complete. The same result holds with "compact" instead of "complete".

7.3. Solutions to Exercises

SOLUTION 7.3.1.

(1) $\mathbb{Q}$ is not complete as it is not closed (or, for instance, the sequence $\left(1 + \frac{1}{n}\right)^n$ belongs to $\mathbb{Q}$ and it is Cauchy in $\mathbb{R}$ and hence Cauchy in $\mathbb{Q}$ but its limit is $e \notin \mathbb{Q}$).

(2) It is not complete as it is not closed. Another way of seeing this is the following: Consider the sequence (x_n) defined by

$$\begin{cases} x_{n+1} = \sqrt{2 + x_n}, & n \in \mathbb{N}, \\ x_1 = \sqrt{2}. \end{cases}$$

Then, it can easily be established that (x_n) is an increasing and bounded above. Hence it converges (and thus Cauchy!). It can also be shown that (x_n) takes its values in $\mathbb{R} \setminus \mathbb{Q}$ and that $x_n \to 2 \notin \mathbb{R} \setminus \mathbb{Q}$.

(3) The set $\mathbb{Q} \cap [3, 4]$ is not complete since it is not closed as we can show that its closure in $\mathbb{R}$ is $[3, 4]$ (cf. Exercise 3.3.19).

(4) The set $(0, \infty)$ is not complete since it is not closed.

(5) The set $\{n : n \geq 1\}$ (which is nothing but $\mathbb{N}$!) is closed in $\mathbb{R}$ since its complement is an infinite union of open sets (hence it is open) and thus $\{n : n \geq 1\}$ is complete.

(6) Again, $\{(-1)^n : n \geq 1\}$ is closed (why?) in $\mathbb{R}$ and hence complete.

(7) $\{\frac{1}{n} : n \in \mathbb{N}\} \cup \{0\}$ is complete (why?).

(8) The given set can be written as

$$\{(x, y) \in \mathbb{R}^2 : x \geq 1, y(x - 1) \geq 1\}$$

Adopting the method of Exercise 4.3.33, we can show that it is closed in $\mathbb{R}^2$, which is complete, yielding the completeness of the given set.

SOLUTION 7.3.2. Let $(x_n)_n$ be a Cauchy sequence in X. Hence for any $\varepsilon > 0$ there exists $N \in \mathbb{N}$ such that for all $n, m \in \mathbb{N}$

$$n, m \geq N \Rightarrow d(x_n, x_m) < \varepsilon.$$

Take for instance $\varepsilon = \frac{1}{4}$, then this gives us $\forall n, m \geq N : x_n = x_m$ which is an eventually constant sequence and in a discrete metric space the only convergent sequences are the eventually constant ones. Thus X is complete.

SOLUTION 7.3.3. Let (x_n) be a Cauchy sequence in X and let $(x_{n(k)})$ be a convergent subsequence to x. Let $\varepsilon > 0$. Since (x_n) is Cauchy, there is some integer N such that

$$\forall n, m \geq N : d(x_n, x_m) < \frac{\varepsilon}{2}.$$

Since $(x_{n(k)})$ converges to x, there is some integer K such that

$$\forall k \geq K : d(x_{n(k)}, x) < \frac{\varepsilon}{2}.$$

For any $n \geq N$, choose $k \geq K$ verifying $n(k) \geq N$ so that we have

$$d(x_n, x) \leq d(x_n, x_{n(k)}) + d(x_{n(k)}, x) < \frac{\varepsilon}{2} + \frac{\varepsilon}{2} = \varepsilon.$$

The proof is complete.

SOLUTION 7.3.4. Let (x_n) be a Cauchy sequence in $\mathbb{R}$. This implies that it is bounded. The Bolzano-Weierstrass theorem implies that (x_n) has a convergent subsequence. But, a Cauchy sequence having a convergent subsequence is itself convergent to the same limit (see Exercise 7.3.3). Therefore, $(\mathbb{R}, |\cdot|)$ is complete.

SOLUTION 7.3.5. Let $(x_n)_n$ be a Cauchy sequence in $(\mathbb{C}, |\cdot|_{\mathbb{C}})$. Since x_n is complex-valued, one can write $x_n = a_n + ib_n$ (where of course $(a_n)_n$ and $(b_n)_n$ are two real-valued sequences). Since $(x_n)_n$ is Cauchy,

$$\forall \varepsilon > 0, \exists N \in \mathbb{N}, \forall m, n \, (m, n \geq N \Longrightarrow |x_n - x_m|_{\mathbb{C}} < \varepsilon).$$

Now both $(a_n)_n$ and $(b_n)_n$ are Cauchy as

$$|a_n - a_m|_{\mathbb{R}} = |\operatorname{Re}(x_n - x_m)|_{\mathbb{R}} \leq |x_n - x_m|_{\mathbb{C}}$$

and

$$|b_n - b_m|_{\mathbb{R}} = |\operatorname{Im}(x_n - x_m)|_{\mathbb{R}} \leq |x_n - x_m|_{\mathbb{C}}.$$

Since $(\mathbb{R}, |\cdot|)$ is complete, both $(a_n)_n$ and $(b_n)_n$ are convergent to a and b respectively. We set $x = a + ib$. Then obviously $x \in \mathbb{C}$ and we need only show that $x_n \to x$ in $(\mathbb{C}, |\cdot|_{\mathbb{C}})$. This follows easily from

$$0 \leq |x_n - x|_{\mathbb{C}} = |(a_n - a) + i(b_n - b)|_{\mathbb{C}} \leq |a_n - a|_{\mathbb{R}} + |b_n - b|_{\mathbb{R}} \to 0.$$

SOLUTION 7.3.6.

(1) Call this space X. Let (x_n) be a Cauchy sequence in X. Since X is compact, it is sequentially compact (we are in a metric space!) so that (x_n) has a subsequence $(x_{n(k)})$ which converges to a point $x \in X$, say. Thus, by Exercise 7.3.3, (x_n) too converges to $x \in X$, proving the completeness of X.

(2) The answer is no in general. For instance, usual $\mathbb{R}$ is complete but not compact (remember that we need also total boundedness for the converse to hold).

(3) By the first answer, we know that for instance $[-2, 2]$ is complete. To prove $\mathbb{R}$ is complete, let (x_n) be a Cauchy sequence in $\mathbb{R}$. Then for all $\varepsilon > 0$ and in particular for $\varepsilon = 2$,

$$\exists N \in \mathbb{N} : \forall n, m \in \mathbb{N} \ (n, m \geq N \Rightarrow |x_n - x_m| \leq 2).$$

Set $y_n = x_{N+n} - x_N$ for all n. Then (y_n) is Cauchy in $[-2, 2]$ and hence it converges, call y its limit. Hence

$$y = \lim_{n \to \infty} y_n = \lim_{n \to \infty} x_n - x_N \implies \lim_{n \to \infty} x_n = y + x_N \in \mathbb{R},$$

proving the completeness of $\mathbb{R}$.

(4) None of the implication is true. For example, in the usual metric $(-1, 1)$ is locally compact but it is not complete. Conversely, the space $C([0, 1], \mathbb{R})$ (see Exercise 7.3.13) is complete but not locally compact (see Exercise 5.3.20).

SOLUTION 7.3.7. Let us show that $(x_n = n)_n$ is a Cauchy sequence in $(\mathbb{R}, d)$ not converging to any point in $\mathbb{R}$. We have

$$d(n, m) = \left| \frac{n}{1 + |n|} - \frac{m}{1 + |m|} \right| = \left| \frac{n}{1 + n} - \frac{m}{1 + m} \right| = \left| \frac{n - m}{(1 + n)(1 + m)} \right|.$$

WLOG, we may assume that $n \geq m$. Hence

$$0 \leq d(n, m) = \left| \frac{n - m}{(1 + n)(1 + m)} \right| \leq \frac{n - m}{nm} \leq \frac{n}{nm} = \frac{1}{m} \longrightarrow 0$$

as $n, m \longrightarrow \infty$.

Now if $(n)_n$ converged to some $a \in \mathbb{R}$ with respect to d, then we would have

$$0 = \lim_{n \to \infty} \left| \frac{n}{1 + |n|} - \frac{a}{1 + |a|} \right| = \left| 1 - \frac{a}{1 + |a|} \right|$$

and it is plain (check it out!) that the equation $\left|1 - \frac{a}{1+|a|}\right| = 0$ has no solution. Thus $(\mathbb{R}, d)$ is not complete.

SOLUTION 7.3.8. Recall the d was defined for every $x, y \in \mathbb{N}$, by

$$d(x, y) = \begin{cases} 0, & x = y, \\ 5 + \frac{1}{x} + \frac{1}{y}, & x \neq y. \end{cases}$$

Now, we proceed to show the completeness of $(\mathbb{N}, d)$. Let (x_n) be a Cauchy sequence in $(\mathbb{N}, d)$. Then

$$\forall \varepsilon > 0, \exists N \in \mathbb{N}, \forall n, m \in \mathbb{N} : (n, m \geq N \Rightarrow d(x_n, x_m) < \varepsilon)$$

or

$$\forall \varepsilon > 0, \exists N \in \mathbb{N}, \forall n, m \in \mathbb{N} : \left(n, m \geq N \Rightarrow 5 + \frac{1}{x_n} + \frac{1}{x_m} < \varepsilon\right)$$

Since this is true for all $\varepsilon > 0$, it is true for $\varepsilon = 5$, say. This gives us a contradiction unless we take $d(x_n, x_m) = 0$. Hence $x_n = x_m$ for all $n, m \geq N$. This is an eventually constant sequence and hence it converges.

SOLUTION 7.3.9.

(1) Let $x = (x_n) = (1, 1, \cdots, 1, \cdots)$ and
$y = (y_n) = (1, \frac{1}{\sqrt{2}}, \frac{1}{\sqrt{3}}, \cdots, \frac{1}{\sqrt{n}}, \cdots)$. Then

$$\sum_{n=1}^{\infty} |x_n|^2 = 1^2 + 1^2 + \cdots + 1^2 + \cdots = +\infty$$

and

$$\sum_{n=1}^{\infty} |y_n|^2 = \sum_{n=1}^{\infty} \left(\frac{1}{\sqrt{n}}\right)^2 = \sum_{n=1}^{\infty} \frac{1}{n} = +\infty$$

since both series are divergent. This means that neither x nor y belongs to ℓ^2.

Now let $x' = (x'_n) = (1, 1, \cdots, 1, 0, \cdots)$ and $y' = (y'_n) = (1, \frac{1}{2}, \frac{1}{3}, \cdots, \frac{1}{n}, \cdots)$. Then both x' and y' belong to ℓ^2 as

$$\sum_{n=1}^{\infty} |x'_n|^2 = 1^2 + 1^2 + \cdots + 1^2 + 0 + \cdots < +\infty$$

and

$$\sum_{n=1}^{\infty} |y'_n|^2 = \sum_{n=1}^{\infty} \left(\frac{1}{n}\right)^2 = \sum_{n=1}^{\infty} \frac{1}{n^2} < +\infty,$$

i.e. both of the involved series converge.

(2) We first prove that d is a metric. We start by showing that d is well-defined (i.e. the series involved in the definition of d converges). We have for all k

$$|x_k - y_k|^2 = |x_k|^2 + 2\operatorname{Re}(x_k y_k) + |y_k|^2 \leq |x_k|^2 + 2|\operatorname{Re}(x_k y_k)| + |y_k|^2$$

Hence

$$|x_k - y_k|^2 \leq |x_k|^2 + 2|x_k y_k| + |y_k|^2$$

for every k. The Cauchy-Schwarz inequality then gives us

$$\sum_{k=1}^{n} |x_k - y_k|^2 \leq \sum_{k=1}^{n} |x_k|^2 + 2\sqrt{\sum_{k=1}^{n} |x_k|^2}\sqrt{\sum_{k=1}^{n} |y_k|^2} + \sum_{k=1}^{n} |y_k|^2.$$

Therefore

$$\sum_{k=1}^{n} |x_k - y_k|^2 \leq \sum_{k=1}^{\infty} |x_k|^2 + 2\sqrt{\sum_{k=1}^{\infty} |x_k|^2}\sqrt{\sum_{k=1}^{\infty} |y_k|^2} + \sum_{k=1}^{\infty} |y_k|^2 < +\infty$$

as $x, y \in \ell^2$. Thus the partial sum $\displaystyle\sum_{k=1}^{n} |x_k - y_k|^2$ is bounded above and increasing since all terms are positive. Thus the series $\displaystyle\sum_{k\geq 1} |x_k - y_k|^2$ converges.

Let us now verify that d satisfies the properties of a metric.

(a) Let $x, y \in \ell^2$.

    (i) Obviously $x = y \Rightarrow d(x, y) = 0$.

    (ii) We have

$$d(x, y) = 0 \Leftrightarrow \sum_{n=1}^{\infty} |x_n - y_n|^2 = 0 \Rightarrow |x_n - y_n|^2 = 0, \ \forall n \Rightarrow x_n = y_n, \ \forall n.$$

Hence $x = y$.

(b) For any $x, y \in \ell^2$, $d(x, y) = d(y, x)$.

(c) Let $x, y, z \in \ell^2$. From the Minkowski inequality with $p = 2$ (see Exercise 2.3.9) one has

$$\sqrt{\sum_{k=1}^{n} |x_k - z_k|^2} \leq \sqrt{\sum_{k=1}^{n} |x_k - y_k|^2} + \sqrt{\sum_{k=1}^{n} |y_k - z_k|^2}$$

and hence

$$\sqrt{\sum_{k=1}^{n} |x_k - z_k|^2} \leq \sqrt{\sum_{k=1}^{\infty} |x_k - y_k|^2} + \sqrt{\sum_{k=1}^{\infty} |y_k - z_k|^2}.$$

A similar argument to one used just above shows that

$$\sqrt{\sum_{k=1}^{\infty}|x_k - z_k|^2} \leq \sqrt{\sum_{k=1}^{\infty}|x_k - y_k|^2} + \sqrt{\sum_{k=1}^{\infty}|y_k - z_k|^2},$$

establishing the triangle inequality for d.

Now we show that (ℓ^2, d) is complete. Let $(x^p)_p$ be a Cauchy sequence in ℓ^2 where $x^p = (x_{1p}, x_{2p}, \cdots, x_{np}, \cdots)$. Let $\varepsilon > 0$, then there exists an $N \in \mathbb{N}$ such that for all $p, q \in \mathbb{N}$ ($p, q \geq N \Rightarrow d(x^p, x^q) < \varepsilon$). Thus for any n we have

$$\sum_{k=1}^{n}|x_{kp} - x_{kq}|^2 \leq d(x^p, x^q) < \varepsilon^2.$$

Taking $q \to \infty$ (and keeping n and p fixed) gives us

$$\sum_{k=1}^{n}|x_{kp} - x_k|^2 \leq \varepsilon^2$$

and hence $\sum_{k=1}^{n}|x_{kp} - x_k|^2$ converges to a limit which is less than or equal to ε^2. Thus the sequence $y_k = x_{kp} - x_k$ belongs to ℓ^2. But $(x^p)_p \in \ell^2$ and hence $(x_k)_k \in \ell^2$. In the end one has $d(x^p, x) \leq \varepsilon$ for all $p \geq N$.

SOLUTION 7.3.10. We first show that (x_n) is Cauchy. Let $\varepsilon > 0$. We proceed as follows (what we will do in the next line will seldom be useful but it really helps here)

$$d(x_n, x_m) = |e^{-n} - e^{-m}| \leq |e^{-n}| + |e^{-m}| = e^{-n} + e^{-m}.$$

One can choose N such that $e^{-N} < \frac{\varepsilon}{2}$ (why?). Then for all $n, m \geq N$, one has

$$e^{-n} + e^{-m} \leq e^{-N} + e^{-N} < \frac{\varepsilon}{2} + \frac{\varepsilon}{2} = \varepsilon,$$

i.e. (x_n) is Cauchy.

We digress a bit to say: **Do not** think that $(-n)$ has $-\infty$ as its limit!! This is not the usual metric!

Now, let us show that (x_n) does not converge in $(\mathbb{R}, d)$. Assume it were and let x be its limit. Then we would have on the one hand

$$|e^{-n} - e^x| \to 0.$$

On the other hand, we know that

$$\lim_{n \to \infty} |e^{-n} - e^x| = e^x.$$

Hence we ended up with $e^x = 0$ which of course does not hold for any $x \in \mathbb{R}$. Thus the limit of the sequence does not exist and (X, d) is not complete.

SOLUTION 7.3.11. The given set A is not complete as it is not closed. Let us show that. Consider the sequence

$$(x_n)_n = \left(1, \frac{1}{2}, \frac{1}{3}, \cdots, \frac{1}{n}, 0, 0, \cdots\right).$$

Then $(x_n)_n$ belongs to A. It has as limit with respect to d the element

$$x = \left(1, \frac{1}{2}, \frac{1}{3}, \cdots, \frac{1}{n}, \frac{1}{n+1}, \frac{1}{n+2}, \cdots\right)$$

as

$$d(x_n, x)^2 = \sum_{k=n+1}^{\infty} \frac{1}{k^2} \to 0 \text{ as } n \to \infty,$$

since this is a remainder of a convergent series.

Evidently, $x \notin A$ and hence A is not complete.

REMARK. We ask the reader the following question which he/she should try to give an answer to. We could (couldn't we?) have taken as "another limit"

$$y = \left(1, \frac{1}{2}, \frac{1}{3}, \cdots, \frac{1}{n}, \frac{1}{2(n+1)}, \frac{1}{2(n+2)}, \cdots\right) \notin A$$

since $d(x_n, y) \to 0$, as $n \to \infty$. We can then have two limits for a convergent sequence in a metric space!! What is then wrong with this reasoning?

SOLUTION 7.3.12.

(1) We show (f_n) is a Cauchy sequence. We can show the "Cauchyness" of (f_n) either by doing some arithmetic or geometrically by observing that $d(f_n, f_{n+p})$ is actually the area of the triangle of apexes $\left(\frac{1}{2} - \frac{1}{n}, 0\right)$, $\left(\frac{1}{2} - \frac{1}{n+p}\right)$ and $\left(\frac{1}{2}, 0\right)$ where $p, n \in \mathbb{N}$. Hence

$$d(f_n, f_{n+p}) = \frac{1}{2} \times 1 \times \left[\frac{1}{2} - \frac{1}{n+p} - \left(\frac{1}{2} - \frac{1}{n}\right)\right]$$

$$= \frac{1}{2}\left(\frac{1}{n} - \frac{1}{n+p}\right) \le \frac{1}{2n}.$$

Hence

$$\forall \varepsilon > 0, \exists N = \left[\frac{\varepsilon}{2}\right] + 1 \in \mathbb{N}, \forall n, p \in \mathbb{N} \ (n \ge N \Rightarrow d(f_n, f_{n+p}) < \varepsilon),$$

establishing the "Cauchyness" of (f_n).

(2) The pointwise limit of (f_n) is

$$f(x) = \begin{cases} 0, & 0 \le x < \frac{1}{2}, \\ 1, & \frac{1}{2} \le x \le 1. \end{cases}$$

Hence

$$d(f_n, f) = \int_0^1 |f_n(x) - f(x)| dx$$

$$= \int_0^{\frac{1}{2} - \frac{1}{n}} 0 dx + \int_{\frac{1}{2} - \frac{1}{n}}^{\frac{1}{2}} \left[nx + \left(1 - \frac{1}{2}n \right) \right] dx + \int_{\frac{1}{2}}^1 0 dx$$

$$= \frac{1}{2n} \to 0 \text{ as } n \to \infty.$$

Now, to show (X, d) is not complete, assume that there is a continuous function g on $[0, 1]$ such that $d(f_n, g) \to 0$ and we will reach a contradiction. Since

$$|f(x) - g(x)| \le |f(x) - f_n(x)| + |f_n(x) - g(x)|, \; \forall x \in [0, 1],$$

we have

$$0 \le \int_0^1 |f(x) - g(x)| dx \le \int_0^1 |f(x) - f_n(x)| dx + \int_0^1 |f_n(x) - g(x)| dx.$$

Passing to the limit and using our hypotheses yield

$$\int_0^1 |f(x) - g(x)| dx = 0$$

and hence $f(x) = g(x)$ at each x for which $f - g$ is continuous, i.e.

$$f(x) = g(x), \; \forall x \in [0, 1] \setminus \left\{ \frac{1}{2} \right\}.$$

But, g is assumed to be continuous on $[0, 1]$ and hence it must be continuous at $\frac{1}{2}$ and we would have

$$\lim_{x \to \frac{1}{2}^-} g(x) = \lim_{x \to \frac{1}{2}^+} g(x) = g \left(\frac{1}{2} \right).$$

However,

$$\lim_{x \to \frac{1}{2}^-} g(x) = \lim_{x \to \frac{1}{2}^-} f(x) = 0 \ne \lim_{x \to \frac{1}{2}^+} g(x) = \lim_{x \to \frac{1}{2}^+} f(x) = 1.$$

Therefore, such a continuous function g cannot exist. In fact, such a function (without continuity) does not even exist and hence (X, d) is not a complete metric space. The solution is over.

SOLUTION 7.3.13. Let (f_n) be a Cauchy sequence in (X, d_∞). Then

$$\forall \varepsilon > 0, \exists N \in \mathbb{N}, \forall n, m \in \mathbb{N} \ (n, m \geq N \Rightarrow d_\infty(f_n, f_m) < \varepsilon)$$

or

$$\forall \varepsilon > 0, \exists N \in \mathbb{N}, \forall n, m \in \mathbb{N} \ (n, m \geq N \Rightarrow \sup_{x \in [0,1]} |f_n(x) - f_m(x)| < \varepsilon).$$

This implies that for each particular x in $[0, 1]$ one has

$$|f_n(x) - f_m(x)| < \varepsilon \text{ for all } n, m \geq N.$$

Thus for each $x \in [0, 1]$, $(f_n(x))_n$ is a Cauchy sequence in usual $\mathbb{R}$. Since the latter is complete, we deduce that $(f_n(x))_n$ must converge to some $f(x) \in \mathbb{R}$ (for each $x \in [0, 1]$) . Thus we have a function f defined for all $x \in [0, 1]$. Fixing n in the last displayed equation, and letting m tend to infinity yield

$$|f_n(x) - f(x)| < \varepsilon \text{ whenever } n \geq N$$

and for each x where N is independent of x. Thus

$$d_\infty(f_n, f) = \sup_{x \in [0,1]} |f_n(x) - f(x)| < \varepsilon \text{ for all } n \geq N.$$

Therefore, we have shown that (f_n) converges in the supremum metric to f, i.e. that (f_n) converges uniformly to f. From the hint, f must therefore be continuous.

In the end, the *arbitrary* Cauchy sequence (f_n) converges in (X, d_∞) to a *continuous* function f. In other words, (X, d_∞) is complete. The proof is over.

SOLUTION 7.3.14. Let (f_n) be a Cauchy sequence in X. As in the previous exercise, we can easily obtain that $(f_n(x))_n$ converges to a function f defined on $[0, 1]$. To finish the proof, two claims are to be shown.

(1) **Claim:** *f is bounded.* Since (f_n) is Cauchy, for $\varepsilon = 1 > 0$, there is some $N \in \mathbb{N}$ such that for all $n, m \geq N$ we have $d(f_n, f_m) < 1$. Hence for $m = N$, we have $d(f_n, f_N) < 1$ whenever $n \geq N$. Remember that f_N is bounded and hence for some $M \geq 0$, $|f_N(x)| \leq M$ for all $x \in [0, 1]$. Now we have

$$\forall x \in [0, 1]: \ |f_n(x)| \leq |f_n(x) - f_N(x)| + |f_N(x)| < 1 + M$$

whenever $n \geq N$. Passing to the limit as n tends to ∞ gives

$$\forall x \in [0, 1]: \ \lim_{n \to \infty} |f_n(x)| = |\lim_{n \to \infty} f_n(x)| = |f(x)| \leq 1 + M,$$

proving the boundedness of f.

(2) **Claim:** $\lim\limits_{n\to\infty} d(f_n, f) \to 0$. Since (f_n) is Cauchy, we know that

$$\forall \varepsilon > 0, \exists N \in \mathbb{N}, \forall n, m \in \mathbb{N} \ (n, m \geq N \Rightarrow d(f_n, f_m) < \varepsilon).$$

But d is the supremum metric, and hence

$$|f_n(x) - f_m(x)| < \varepsilon, \ \forall x \in [0, 1]$$

and whenever $m, n \geq N$. Keeping n fixed and taking the limit as m goes to infinity yield

$$|f_n(x) - f(x)| \leq \varepsilon$$

whenever $n \geq N$ and for all $x \in [0, 1]$. Thus we are led to

$$d(f_n, f) \leq \varepsilon$$

for $n \geq N$. That is $\lim\limits_{n\to\infty} d(f_n, f) \to 0$.

The proof is complete.

SOLUTION 7.3.15.

(1) To show that $x \mapsto e^x$ is not a polynomial we use a contradiction argument. Assume that $x \mapsto e^x$ is a polynomial of degree m, say, i.e. for some real $a_0, a_1, \cdots, a_m$ we have

$$e^x = a_0 + a_1 x + \cdots + a_m x^m.$$

It is known from the course of calculus that $x \mapsto e^x$ is C^{m+1} (it is in fact C^∞). Hence by differentiating $(m+1)$-times both sides of the previous displayed equation we get

$$(e^x)^{(m+1)} = e^x = (a_0 + a_1 x + \cdots + a_m x^m)^{(m+1)} = 0$$

which is of course absurd. Thus the exponential function cannot be a polynomial.

(2) Of course $P_n(x) = \left(1 + \frac{x}{n}\right)^n$ (for $n \geq 1$) is a polynomial sequence ($\sum\limits_{n=0}^{N} \dfrac{x^n}{n!}$ is another polynomial sequence converging to e^x). So, it only remains to show that $d(P_n, e^x) \to 0$ as $n \to \infty$ and by uniqueness of the limit e^x is the only limit of (P_n) (see the remark below for a comment on this). Let us show now $P_n \to e^x$ in (X, d). To this end, we calculate $\sup\limits_{x \in [0,1]} |P_n(x) - e^x|$.

We can write

$$\sup_{x \in [0,1]} |P_n(x) - e^x| \leq e^1 \sup_{x \in [0,1]} |P_n(x) e^{-x} - 1|.$$

Studying the function $P_n(x)e^{-x} - 1$ on $[0,1]$ allows us to write

$$0 \leq |P_n(x)e^{-x} - 1| \leq 1 - P_n(1)e^{-1} = 1 - \left(1 + \frac{1}{n}\right)^n e^{-1}.$$

Taking the supremum and passing to the limit as $n \to \infty$ yield

$$0 \leq d(P_n, e^x) \leq e\left(1 - \left(1 + \frac{1}{n}\right)^n e^{-1}\right) \to 0.$$

REMARK. We would like to give a comment on this answer. We did not consider this reasoning, i.e. the use of the uniqueness of the limit, as a correct one in the True or False section for the "integral metric".

The reason here differs from the other one. The minimum one suggests for the integral metric to be well-defined is that the functions be Riemann-integrable (in such case the limit is not unique as we do not have a distance) whilst in this case (the supremum metric) what one suggests is that at least the functions be bounded and d in this case remains a metric. Hence the limit is unique.

(3) Take any continuous function f which is not a polynomial (e^x, $\sin x$,...etc) and hence $f \notin X$. By the well-known Weierstrass theorem (see also the next chapter) we know that there is a polynomial sequence $(P_n)_n$ converging uniformly to f, i.e. $P_n \to f$ with respect to d. Hence $(P_n)_n$ is Cauchy but converging to f which is outside X. Thus (X, d) is not complete.

SOLUTION 7.3.16.

(1) This a simple consequence of Test 5.

(2) A is not closed and hence it is not closed.

(3) Let (x_n) be a Cauchy sequence in (A, d'), i.e.

$$\forall \varepsilon > 0, \exists N \in \mathbb{N}, \forall n, m \in \mathbb{N} \ (n, m \geq N \Rightarrow d'(x_n, x_m) = |\ln x_n - \ln x_m| < \varepsilon).$$

We immediately see that $(\ln x_n)$ becomes a Cauchy sequence in usual $\mathbb{R}$ which is obviously complete. Thus there exists some $y \in \mathbb{R}$ such that $|\ln x_n - y| \to 0$ as $n \to \infty$. Set $x = e^y$. Then $x > 0$, i.e. $x \in A$. Besides,

$$d'(x_n, x) = |\ln x_n - \ln x| = |\ln x_n - y| \to 0 \text{ as } n \to \infty.$$

(4) We show that $id : (A, d) \to (A, d')$ is a homeomorphism. The bijectivity of id is evident. It is continuous since the logarithm function is continuous on $(0, \infty)$ (from basic real analysis!). The inverse function $id^{-1} = id : (A, d') \to (A, d)$ is also continuous as the exponential function is in $[0, \infty)$.

(5) We deduce from the previous answers that completeness is not a topological property.

SOLUTION 7.3.17.

(1) Let f be such a function. By the mean value theorem applied to f on $[x, y]$ $(x, y \in [a, b])$ we have

$$\exists c \in (x, y) : \ f(y) - f(x) = (y - x)f'(c)$$

and by the hypothesis on the derivative of f

$$|f(y) - f(x)| \le M|y - x|,$$

i.e. f is a contraction.

(2) Transform the given equation to

$$\frac{1}{32}x^4 + \frac{x^3}{2} - x - \frac{1}{4} = 0 \text{ or } \frac{1}{32}x^4 + \frac{x^3}{2} - \frac{1}{4} = x.$$

Set $f(x) = \frac{1}{32}x^4 + \frac{x^3}{2} - \frac{1}{4}$. Then it can easily be shown that

$$|f'(x)| \le \frac{25}{64} < 1 \text{ for all } x \in \left[0, \frac{1}{2}\right].$$

Thus the given equation has a root between 0 and $\frac{1}{2}$.

SOLUTION 7.3.18.

(1) Set $g = f^n$. Since g is a contraction, there exists a unique $x \in X$ such that $g(x) = x$. Now

$$\begin{aligned} d(x, f(x)) &= d(g(x), f(x)) = d(g(x), (f \circ g)(x)) \\ &= d(g(x), (g \circ f)(x)) \\ &\le Md(x, f(x)) \text{ since } g \text{ is a a contraction.} \end{aligned}$$

Since $M < 1$, we immediately see that $f(x) = x$. Thus, if f^n is a contraction (then it admits a unique fixed point), then f has the same fixed point. To show its uniqueness, assume y is another fixed point of f. Then y is necessarily a fixed point for f^n for

$$f^n(y) = f^{n-1}(f(y)) = f^{n-1}(y) = \cdots = f(y) = y.$$

But the fixed point of f^n is unique and hence $x = y$ and thus the proof is over.

(2) • Define $\cos x : [0, \frac{\pi}{2}] \to [0, 1]$. Then this function cannot be a contraction since $\frac{|\cos x - 0|}{x - \frac{\pi}{2}}$ tends to 1 as x tends to $\frac{\pi}{2}$. Hence It cannot be bounded above by some $k < 1!$.

REMARK. Although $\cos x$ is not a contraction, $\alpha \cos x$ is always one whenever $0 \leq \alpha < 1$.

Let $g(x) = \cos(\cos x)$. Then $f'(x) = \sin x \sin(\cos x)$. Then

$$\sup_{0 \leq x \leq \frac{\pi}{2}} f'(x) \leq \sin(\cos x) \leq \sin 1 < 1.$$

Thus the mean value theorem gives for all x and y

$$|\cos^2 x - \cos^2 x| \leq \sin 1 |x - y|,$$

i.e. $\cos^2$ is a contraction and so by the first question $\cos x$ has a unique fixed point.

- The way of finding the fixed point is well described in the proof of the fixed point theorem. We start with a point x_0. Then we calculate successively $x_2 = \cos x_1$, $x_3 = \cos x_2,$...etc. If we start with $x_1 = 0.7$ in radians and not degrees! We already knew roughly where the root would be graphically! Otherwise we could have started by any value in $(0, 1)$. Then

$$x_2 = \cos(x_1) = 0.7648,$$
$$x_3 = \cos(x_2) = 0.7215,$$
$$x_4 = \cos(x_3) = 0.7508 \cdots$$

and we carry on until this process stabilizes at $x_{18} = 0.7391$, accurate to four decimal places (this is a little slow and other methods give here a better result like Newton's).

SOLUTION 7.3.19.

(1) Let $x, y \in X$. Then $x, y \geq 1$ and hence

$$|f(x) - f(y)| = |x - y| \left| \frac{1}{2} - \frac{1}{xy} \right|.$$

But

$$0 < \frac{1}{xy} \leq 1 \text{ and so } \left| \frac{1}{2} - \frac{1}{xy} \right| \leq \frac{1}{2}.$$

Therefore,

$$\forall x, y \in X : \ |f(x) - f(y)| \leq \frac{1}{2} |x - y|.$$

(2) Assume there is $x \in X$ such that $f(x) = x$. Then

$$\frac{x}{2} + \frac{1}{x} = x \text{ or } x^2 = 2$$

and the previous equation clearly does not hold for any $x \in X$ since $\pm\sqrt{2} \notin \mathbb{Q}$!

(3) It does not contradict the fixed point theorem since X is not complete (why?) even if f is a contraction. Therefore, the fixed point is not guaranteed to exist anymore. This tells us that the hypothesis on the space being complete cannot be merely dropped.

SOLUTION 7.3.20.

(1) Let $x, y \geq 1$. We could use similar estimates to those in the previous exercise, but it is important here to show that 1 is the best constant. To achieve this aim, we use the mean value theorem, which, applied to f on $[x, y]$ gives

$$\exists c \in (x, y) : \; f(x) - f(y) = f'(c)(x - y).$$

We now need to check that $\sup_{x \geq 1} |f'(x)| = 1$ which we leave to the reader. Hence

$$\forall x, y \geq 1 : \; |f(x) - f(y)| < |x - y|.$$

(2) It is plain that

$$\forall x \geq 1 : \; f(x) \neq x.$$

(3) It does not contradict the fixed point theorem since f is not a contraction despite the fact that X is complete (hence the fixed point is not guaranteed to exist anymore). This shows the importance of the contraction hypothesis.

This example does not contradict the result of Exercise 5.3.27 either because $[1, \infty)$ is not compact.

SOLUTION 7.3.21.

(1) Suppose that (X', d') is complete and let us show that (X, d) is in its turn complete. Let (x_n) be a Cauchy sequence in X. Then $d(x_n, x_m) \to 0$ as $n, m \to \infty$ and so

$$\lim_{n,m \to \infty} d'(f(x_n), f(x_m)) = 0.$$

This means that $(f(x_n))$ is Cauchy in X', assumed complete, and hence it must converge to some $y \in X' = f(X)$. But there exists one $x \in X$ such that $y = f(x)$. Thus

$$d'(f(x_n), y) = d'(f(x_n), f(x)) = d(x_n, x) \to 0$$

as n goes to infinity. Thus (X, d) is complete.

(2) The proof of the other implication is very similar to the previous one and we leave it to the reader.

SOLUTION 7.3.22. Both examples are in the setting of $\mathbb{R}$ endowed with the usual topology ($\mathbb{R}$ is complete).

(1) Let $A_n = [n, \infty)$. Then each A_n is closed, non-empty and decreasing (with respect to $\subset$). However, we see that $\bigcap_{n=1}^{\infty} A_n = \varnothing$ which is not a contradiction for $d(A_n) \not\to 0$ as $n \to \infty$.

(2) Consider $A_n = (1, 1 + \frac{1}{n}]$. Then (A_n) is decreasing and verifies $d(A_n) \to 0$ as n goes to infinity. However, $\bigcap_{n=1}^{\infty} A_n = \varnothing$. This is also not a contradiction for A_n is not closed.

SOLUTION 7.3.23. Recall that a metric space is compact iff it is complete and totally bounded. We already know that $[a, b]$ is complete for it is closed in $\mathbb{R}$ which is complete. We need only show that it is totally bounded. Let $\varepsilon > 0$. Choose n such that $\frac{2}{n} \leq \varepsilon$. Consider now the cover $\{B_n\}$ defined as

$$B_0 = \left[a, a + \frac{1}{n}\right), \quad B_i = \left(x_i - \frac{1}{n}, x_i + \frac{1}{n}\right) \text{ and } B_n = \left(b - \frac{1}{n}, b\right].$$

We may easily check that the diameter of B_n, designated by δ, is less than or equal to $\frac{2}{n}$. So $[a, b]$ is totally bounded.

SOLUTION 7.3.24.

(1) Assume $\mathbb{R}$ is countable. Hence it can be written as

$$\mathbb{R} = \{x_1, x_2, \cdots, x_n, \cdots\} = \bigcup_{n=1}^{\infty} C_n \text{ where } C_n = \{x_n\}.$$

In $\mathbb{R}$ the set C_n is closed and $\overset{\circ}{C_n} = \varnothing$. Since $\mathbb{R}$ is complete, Baire's theorem tells us that $\bigcup_{n=1}^{\infty} \overset{\circ}{C_n}$ must be dense but this is clearly not the case as one can clearly see that this union is the empty set. Thus $\mathbb{R}$ is not countable, as required.

(2) As for $\mathbb{R} \setminus \mathbb{Q}$ we know that $\mathbb{Q}$ is countable. Now assume that $\mathbb{R} \setminus \mathbb{Q}$ is also countable. Then $\mathbb{Q} \cup \mathbb{R} \setminus \mathbb{Q} = \mathbb{R}$ would be countable and this contradicts the uncountability of $\mathbb{R}$. Thus $\mathbb{R} \setminus \mathbb{Q}$ is not countable.

SOLUTION 7.3.25. The proof is based on the Baire's theorem which, an equivalent version of it, states that if X is a complete metric space, then the countable intersection of open and dense sets in X is dense in X.

Since C is closed in usual $\mathbb{R}$, it is complete. Assume that C does not have any isolated points, that is, it posses limit points only. Let $x \in C$. Since C is separated, $\{x\}$ is closed in C and so $C \setminus \{x\}$ is open in C. Set $U_x = C \setminus \{x\}$. Since C does not have any isolated point, $\overline{U_x} = C$, that is, U_x is dense in C. According to Baire's theorem, the intersection of U_x $(x \in C)$ must be dense in C. But

$$\bigcap_{x \in C} U_x = \varnothing$$

which is never dense in C. We hence conclude that C does not have any isolated point.

SOLUTION 7.3.26. We need only show that $\overline{f(A)} \subset f(A)$. Let $y \in \overline{f(A)}$, then there exists a sequence (x_n) such that $x_n \in A$ with $y = \lim\limits_{n \to \infty} f(x_n)$. Hence $(f(x_n))_n$ is a Cauchy sequence in Y which is complete. By hypothesis

$$d'(f(x_n), f(x_m)) \geq d(x_n, x_m)$$

for all n, m and so $(x_n)_n$ becomes in its turn a Cauchy sequence but in X.

But A is closed in X and hence it is also complete. This means that $d(x_n, x) \to 0$ with x in A. Now, since f is continuous, one obtains

$$f(A) \ni f(x) = \lim_{n \to \infty} f(x_n) = y,$$

establishing the result.

7.4. Hints/Answers to Tests

SOLUTION 53. No! (Why?)...

SOLUTION 54. Use the sequence defined by $x_n = n$...

SOLUTION 55. Apply Baire's theorem to $[0,1]$...

SOLUTION 56. Use Exercises 4.5.9 & 7.3.25.

CHAPTER 8

Function Spaces

8.2. True or False: Answers

(1) False! We provide a counterexample. Let

$$f_n(x) = \frac{1}{1 + nx}, \quad 0 < x < 1.$$

It can be shown that (f_n) is decreasing. Also, it is plain that

$$\lim_{n \to \infty} \frac{1}{1 + nx} = 0 = f(x)$$

and so f is continuous on $(0, 1)$. Nonetheless,

$$\sup_{0 < x < 1} |f_n(x) - f(x)| = \sup_{0 < x < 1} \left| \frac{1}{1 + nx} \right| = 1 \not\to 0,$$

i.e. the convergence is not uniform.

If, however, the (f_n)s are defined on a *compact* set, then this is true (Dini's theorem).

(2) True! Let $x \leq y$. Since $x \mapsto f_n$ is increasing, $f_n(x) \leq f_n(y)$ for each n. Hence

$$\lim_{n \to \infty} f_n(x) \leq \lim_{n \to \infty} f_n(y) \Longrightarrow f(x) \leq f(y),$$

i.e. f is increasing.

REMARKS.
(a) The result remains true if "decreasing" replaces "increasing" and the proof is very similar.
(b) The result is no longer true if "strictly increasing" replaces "increasing". The reader should try to give a counterexample.

(3) If (f_n) is a convergent sequence of bounded functions, then its pointwise limit f need not be bounded. Let (f_n) be defined as follows

$$f_n(x) = \begin{cases} e^x, & |x| \leq n, \\ 0, & |x| > n. \end{cases}$$

235

Then every (f_n) is bounded (why?). Its pointwise limit, which is clearly seen to be $f(x) = e^x$, is not bounded on $\mathbb{R}$.

But, if the convergence is uniform, then the uniform limit must be bounded. we give a proof although it is similar in core to that of Exercise 7.3.14.

Since (f_n) converges uniformly to f, for all $\varepsilon > 0$ and in particular for $\varepsilon = 1$,

$$\exists N \in \mathbb{N}, \ \forall n \in \mathbb{N}, \ n \geq N \Rightarrow |f_n(x) - f(x)| \leq 1 \ \text{ for all } x \in \mathbb{R}.$$

But all (f_n)s are bounded and in particular f_N is. Whence for some positive C and all $x \in \mathbb{R}$: $|f(x)| \leq C$. Therefore, for each $x \in \mathbb{R}$

$$|f(x)| \leq |f(x) - f_N(x)| + |f_N(x)| \leq 1 + C,$$

i.e. f is bounded.

(4) False! and $f_n(x) = x^n$ on $[0, 1]$ is again a counterexample. Then (f_n) converges pointwise to the nil function on $[0, 1)$. The convergence remains uniform on $[0, a]$ for each $0 < a < 1$ as

$$\lim_{n \to \infty} \sup_{0 \leq x \leq a} |x^n - 0| = \lim_{n \to \infty} a^n = 0$$

and we already saw above the the convergence is not uniform on $[0, 1)$.

(5) False! The following is a counterexample: $f_n(x) = (1 - x)^n$ on $(0, 1]$. Details are left to the reader.

(6) Although continuity implies Riemann integrability, the result does not hold in general (that is, we still need uniform continuity).

As a counterexample, let (f_n) be defined on $[0, 1]$ as follows

$$f_n(x) = \begin{cases} 4n^2x, & 0 \leq x \leq \frac{1}{2n}, \\ -4n^2x + 4n, & \frac{1}{2n} < x < \frac{1}{n}, \\ 0, & \frac{1}{n} \leq x \leq 1. \end{cases}$$

It is clear that (f_n) converges pointwise to $f(x) = 0$ for all $x \in [0, 1]$. Drawing a graph tells us that $\int_0^1 f_n(x)dx$ is actually the area of a triangle of apexes $(0, 0)$, $(\frac{1}{n}, 0)$ and $(\frac{1}{2n}, 2n)$. So

$$\int_0^1 f_n(x)dx = \frac{1}{n} \times (2n) \times \frac{1}{2} = 1.$$

However,

$$\lim_{n \to \infty} \int_0^1 f_n(x)dx = 1 \neq \int_0^1 \lim_{n \to \infty} f_n(x)dx = \int_0^1 f(x)dx = 0.$$

REMARK. In the Lebesgue Integration Course, we do not even need uniform convergence for the result to hold. Many handicaps will then be overcome, in a different context though. This will not be discussed in the present book.

(7) False! Consider $f_n(x) = \frac{x^n}{n}$ defined on $[0,1]$. Then it is evident that f_n converges pointwise to $f(x) = 0$ for every $x \in [0,1]$, though

$$\lim_{n\to\infty} f_n'(x) = \lim_{n\to\infty} x^{n-1} = \begin{cases} 0, & 0 \le x < 1, \\ 1, & x = 1, \end{cases} \ne f'(x) = 0.$$

8.3. Solutions to Exercises

SOLUTION 8.3.1.

(1) It is clear that (f_n) converges pointwise to the zero function (defined on $[0,1]$). Call this limit f.

(2) We can also easily show that (f_n) converges uniformly to the f, i.e. that

$$\lim_{n\to\infty} \sup_{x\in[0,1]} |f_n(x) - f(x)| = 0.$$

SOLUTION 8.3.2.

(1) We have

$$\lim_{n\to\infty} f_n(x) = \lim_{n\to\infty} \frac{e^{nx}}{\sqrt{n}} = 0$$

for every $x \in [-1, 0]$. Hence (f_n) converges to the zero function. Designate this limit by f.

(2) Let us verify that the convergence is uniform. We have

$$\lim_{n\to\infty} \sup_{x\in[-1,0]} |f_n(x) - f(x)| = \lim_{n\to\infty} \sup_{x\in[-1,0]} \frac{e^{nx}}{\sqrt{n}} = \lim_{n\to\infty} \frac{1}{\sqrt{n}} = 0.$$

(3) We obviously have

$$f_n'(x) = \sqrt{n} e^{nx}, \ \forall -1 \le x \le 0$$

which does not converge pointwise on $[-1, 0]$ for $f_n'(0) = \sqrt{n}$. A fortiori, it does not converge uniformly.

REMARK. We may show that (f_n) converges pointwise to $f = 0$ on $[-1, 0)$, but it still does not converge uniformly to the zero function on this interval.

SOLUTION 8.3.3. We apply Dini's theorem. Thus we have to show that (P_n) is monotonic and it converges uniformly to $|x|$. We first claim

$P_n(x) \le |x|$ for all n. To prove it, we use a proof by induction. It is clear $P_0(x) = 0 \le |x|$ (for all x). Assume $P_n(x) \le |x|$. Then

$$|x| - P_{n+1}(x) = |x| - P_n(x) - \frac{1}{2}(|x| - P_n(x))(|x| + P_n(x))$$

$$= \frac{1}{2}(|x| - P_n(x))(2 - |x| - P_n(x)) \ge 0$$

by the induction hypothesis and $x \in [-1, 1]$ hypotheses. We can easily check that $P_n(x) \ge 0$ for all n and all x. To apply Dini's theorem, it only remains to check that (P_n) is monotonic. It is in fact increasing as for all n (and all $x \in [-1, 1]$)

$$P_{n+1}(x) - P_n(x) \ge 0.$$

Therefore, $(P_n(x))$ is an increasing and bounded above sequence and so it converges (pointwise) to some $f(x)$ (with $0 \le f(x) \le |x|$). Finally, it is clear that $f(x)$ verifies $x^2 = f^2(x)$ or $f(x) = |x|$. Thus, all the hypotheses of Dini's theorem are gathered, implying that (P_n) converges uniformly to $|x|$.

SOLUTION 8.3.4. The sequence (f_n) converges pointwise, compactly and uniformly to the function $x \mapsto 0$.

As for (g_n), it converges to $x \mapsto g(x) = x$; pointwise and compactly, but not uniformly.

SOLUTION 8.3.5. To show that V is closed, we take an arbitrary sequence, (x_n) say, in V, i.e. $(f_n(x_n))$ is Cauchy in Y, and we must show that its limit x remains in V, i.e. $(f_n(x))$ is Cauchy in Y.

If p, q, n are positive integers, then we may write

$$d'(f_p(x), f_q(x)) \le d'(f_p(x), f_p(x_n)) + d'(f_p(x_n), f_q(x_n)) + d'(f_q(x), f_q(x)).$$

Let $\varepsilon > 0$. Since (f_n) is continuous at x, for some $\alpha > 0$ and all $n \in \mathbb{N}$ and all $y \in X$

$$d(x, y) < \alpha \implies d'(f_n(x), f_n(y)) < \frac{\varepsilon}{3}.$$

But $d(x_n, x) \to 0$, and hence there exist $n' \in \mathbb{N}$ for which $d(x_{n'}, x) < \alpha$. Therefore,

$$d'(f_q(x_{n'}), f_q(x)) < \frac{\varepsilon}{3} \text{ and } d'(f_p(x_{n'}), f_p(x)) < \frac{\varepsilon}{3}$$

for all $p, q \ge n'$. We also observe that $(f_{n'})$ is Cauchy, so

$$\exists n'' \ge n', \forall p, q \in \mathbb{N} : (p, q \ge n'' \implies d'(f_p(x_{n'}), f_q(x_{n'})) < \frac{\varepsilon}{3}).$$

Thus it becomes clear that

$$\forall \varepsilon > 0, \exists n'' \in \mathbb{N} : \forall p, q \in \mathbb{N} : ((p, q \ge n'' \implies d'(f_p(x), f_q(x)) < \varepsilon),$$

$(f_n(x))$ is Cauchy, i.e. $x \in V$ so that V is closed.

SOLUTION 8.3.6.

(1) Since k is continuous on the compact $[0,1]^2$, it is uniformly continuous. This *implies* that

$$\forall \xi > 0, \exists \alpha > 0, \forall x, y, t \in [a, b] : (|x-t| < \alpha \implies |k(x,y)-k(t,y)| < \xi).$$

But (f_n) is bounded, hence

$$\exists M \geq 0 : \forall x \in [0,1] : |f_n(x)| \leq M.$$

Let x be fixed. Let $\varepsilon > 0$. Set $\xi = \frac{\varepsilon}{M}$. Since k is uniformly continuous, we may find $\alpha > 0$ such that $|x - t| < \alpha$ such that

$$|k(x,y) - k(t,y)| < \xi = \frac{\varepsilon}{M}$$

$$|Kf_n(x) - Kf_n(t)| \leq \int_0^1 |k(x,y) - k(t,y)||f_n(y)|dy$$

$$\leq M \int_0^1 |k(x,y) - k(t,y)|dy$$

$$\leq M \int_0^1 \frac{\varepsilon}{M} dy$$

$$\leq \varepsilon,$$

i.e. (Kf_n) is equicontinuous at each $x \in [0,1]$, hence it is so on $[0,1]$.

(2) Set $(Kf_n) = H$. The set $H(x)$ is bounded for each x. Hence its closure is a closed and bounded set in usual $\mathbb{R}$, so $\overline{H(x)}$ is compact. By Ascoli theorem (Kf_n) is relatively compact.

SOLUTION 8.3.7. Since all the moments are worth zero, so is their sum and hence

$$\int_a^b f(x)p(x)dx = 0$$

where $p(x) = a_n x^n + \cdots + a_1 x + a_0$ is any polynomial with real coefficients. Now, the function f, being continuous on the compact set $[a, b]$, is bounded and hence for some positive C and all $x \in [a, b]$: $|f(x)| \leq C$. By the Weierstrass theorem, for every $\varepsilon > 0$, there exists a polynomial q such that

$$|f(x) - q(x)| < \varepsilon$$

for all $x \in [a, b]$. By the above observation, $\int_a^b f(x)q(x)dx = 0$ too and hence we have

$$
\begin{aligned}
0 \le \int_a^b f^2(x)dx &= \int_a^b (f^2(x) - f(x)q(x))dx \\
&= \int_a^b f(x)(f(x) - q(x))dx \\
&= \left| \int_a^b f(x)(f(x) - q(x))dx \right| \quad \text{(why?)} \\
&\le \int_a^b |f(x)||f(x) - q(x)|dx \\
&< C\varepsilon.
\end{aligned}
$$

Since this is true for all ε, we immediately deduce that $\int_a^b f^2(x)dx = 0$ which, combined with the fact that f^2 is continuous and positive, yield $f^2(x) = 0$ for all $x \in [a, b]$ and thus $f = 0$.

SOLUTION 8.3.8. Set

$$
d_\infty(f, g) = \sup_{x \in [0,1]} |f(x) - g(x)| \text{ for all } f, g \in C([0, 1], \mathbb{R}).
$$

First remember that we have to find a dense and countable set in $C([0, 1], \mathbb{R})$. Let $f \in C([0, 1], \mathbb{R})$. If we come to show that for every $\varepsilon > 0$, there exists a polynomial q with rational coefficients such that $d_\infty(f, g) < \varepsilon$, then Exercise 1.2.4 (which tells us that the set of polynomials with rational coefficients is countable) will allow us to conclude that $C([0, 1], \mathbb{R})$ is separable.

Since f is continuous on $[0, 1]$, there exists -by means of the Weierstrass theorem- a polynomial p such that $d_\infty(f, p) < \varepsilon$. If all the coefficients of p are rational, then we have our polynomial. Otherwise, p will have irrational coefficients (not necessarily all). So if we assume p has degree n, then

$$
p(x) = \alpha_n x^n + \cdots + \alpha_1 x + \alpha_0 = \sum_{i=0}^n \alpha_i x^i
$$

where *not all* α_i ($i = 0, 1, \ldots; n$) are rational. Now, it is clear that every interval centered at α_i with length 2ε contains a rational β_i (and hence $|\alpha_i - \beta_i| < \varepsilon$ for each i). Thus we have recovered a polynomial q with rational coefficients β_i written as

$$
q(x) = \beta_n x^n + \cdots + \beta_1 x + \beta_0 = \sum_{i=0}^n \beta_i x^i.
$$

Therefore, it only remains to approximate $p(x)$ by $q(x)$ (in the supremum distance). We have for all x in $[0,1]$

$$|p(x) - q(x)| \leq \sum_{i=0}^{n} |\beta_i - \alpha_i| x^i < \varepsilon(n+1)$$

which implies that $d_\infty(p,q) < \varepsilon(n+1)$. Thus

$$d_\infty(f,q) \leq d_\infty(f,p) + d_\infty(p,q) < \varepsilon(n+1) + \varepsilon = \varepsilon(n+2)$$

and the proof is complete (those who are superstitious may play with the epsilons in the beginning so that they get one nice epsilon in the end!).

8.4. Hints/Answers to Tests

SOLUTION 57. The sequence (f_n) converges pointwise to $\frac{1}{1-x}$ on $(-1,1)$...The convergence is not uniform on $(-1,1)$ but it is so on $[-a,a]$ where $a < 1$.

SOLUTION 58. The given sequence converges uniformly to the function defined by $f(x) = x$...

SOLUTION 59. f is continuous (why?)...

SOLUTION 60. Yes (prove it!)...

Bibliography

1. C. D. Aliprantis, O. Burkinshaw, *Problems in Real Analysis*, Academic Press, 1999.
2. V. Bryant, *Metric Spaces: iteration and application*, Cambridge University Press, 1985.
3. G. Cohen, *A Course in Modern Analysis and Its Applications*, AMSLS **17**, Cambridge University Press, 2003.
4. J. B. Conway, A Course in Functional Analysis, *Springer*, 1990 (2nd edition).
5. J. Doboš, Sums of closed graph functions. Real functions (Liptovský Jàn, 1996). *Tatra Mt. Math. Publ.*, **14** (1998), 9-11.
6. I. Kaplansky, *Set Theory and Metric Spaces*, AMS Chelsea Publishing, 1977 (second edition), reprinted by the AMS in 2001.
7. S. Lipschutz, *Schaum's Outline of Theory and Problems of General Topology*, Schaum's Outline Series: McGraw-Hill Book Company, 1965.
8. P. E. Long, Classroom Notes: Functions with Closed Graphs, *Amer. Math. Monthly*, **76/8** (1969), 930-932.
9. J. M. Møller, *General Topology*, http://www.math.ku.dk/~moller.
10. J. R. Munkres, *Topology*, Prentice Hall, 2000 (2nd edition).
11. W. Rudin, *Principles of Mathematical Analysis*, McGraw-Hill International Editions, Mathematics Series, 1976 (3rd edition).
12. W. Sierpinski, Un théorème sur les continus, *Tôhoku Math. J.*, **13** (1918) 300-303.
13. Y. Sonntag, *Topologie et Analyse Fonctionnelle* (French), Ellipses, 1998.
14. L. A. Steen, J. A. Seebach, Jr, *Counterexamples in Topology*, Dover, 1995.
15. M. Stoll, *Introduction to Real Analysis*, Addison-Wesley Higher Mathematics, 2001 (2nd edition).
16. W. A. Sutherland, *Introduction to Metric and Topological Spaces*, Oxford University Press, 1975.

Index